# 67 Structure and Bonding

# Coordination Compounds: Synthesis and Medical Application

With Contributions by
R. A. Bulman A. M. J. Fichtinger-Schepman R. Krause
J. Reedijk A. T. van Oosterom P. van de Putte

With 24 Figures and 21 Tables

Springer-Verlag
Berlin Heidelberg NewYork
London Paris Tokyo

ISBN 3-540-17881-3 Springer-Verlag Berlin Heidelberg New York
ISBN 0-387-17881-3 Springer-Verlag New York Berlin Heidelberg

Library of Congress Cataloging-in-Publication Data. Coordination compounds. (Structure and bonding; 67) Bibliography: p. 1. Coordination compounds. I. Bulman, A. (Allen), 1947– .II. Series. QD461.S92 Vol. 67 541.2'2 s 87-16402 [QD474] [541.2'242]
ISBN 0-387-17881-3 (U.S.)



Typesetting: Mitterweger Werksatz GmbH, 6831 Plankstadt, Germany
Printing: Druckhaus Beltz, 6944 Hemsbach, Germany
Bookbinding: J. Schäffer GmbH & Co.KG., 6718 Grünstadt, Germany
2152/3140-543210

# Table of Contents

# Synthesis of Ruthenium(II) Complexes of Aromatic Chelating Heterocycles: Towards the Design of Luminescent Compounds

**Ronald A. Krause**

Department of Chemistry, University of Connecticut, Storrs, CT, 06268, USA

The majority of ruthenium(II) compounds which have been observed to be luminescent are complexes of aromatic heterocyclic chelating ligands. Methods of synthesizing such complexes are reviewed and recommendations made (where considered possible) as to the best procedures.

Models for the emitting excited state are reviewed, particularly with regard to the effects of molecular structure on excited state behavior. Data are tabulated for known emitting complexes. While a wide variety of luminescent ruthenium(II) complexes cannot be designed at present, some guidelines are beginning to emerge.

In the future we look for continued development of the excited state model. Further structure-emission efficiency correlations and correlation with solvent properties are seen as being important.

## Abbreviations Used

In many cases ligands are abbreviated by giving a substituent and its location followed by the ligand abbreviation (e.g., 4,4′-Me-bpy would be 4,4′-dimethyl-2,2′-bipyridyl). In tables collecting complexes of one ligand type only the substituent is given. Ligand structures appear in Fig. 1.

| | |
|---|---|
| Azpy | structure **34** |
| bpy | 2,2′-bipyridyl |
| bpym | structure **23** |
| bpyz | structure **24** |
| bq | structure **42** |
| bt | structure **37** |
| BzImH | structure **6** |
| cod | 1,5-cyclooctadiene |
| CT | charge transfer |
| daf | structure **45** |
| dinapy | structure **32** |
| DMCH | structure **50** |
| DMF | N,N-dimethylformamide |
| DMSO | dimethylsulfoxide |
| dpp | structure **11** |
| dpt | structure **48** |
| dqp | structure **49** |
| en | 1,2-diaminoethane |
| Hdpa | structure **33** |
| Htf | trifluoromethane sulfonic acid |
| i-bq | structure **51** |
| MC | metal centered |
| Me | methyl |
| MLCT | metal-ligand charge transfer |
| NAz | structure **35** |
| nbd | norbornadiene |
| NPP | structure **44** |
| ox | oxalate |
| ph | phenyl |
| phen | 1,10-phenanthroline |
| pq | structure **43** |
| py | pyridine |
| pynapy | structure **31** |
| TAP | structure **30** |
| tf | trifluoromethane sulfonate anion |
| tro | structure **47** |
| trpy | 2,2′,6′,2″-tripyridyl |
| tsite | structure **46** |

Absorbance (energy and molar ε value (E)) and emission spectral data are collected in the following tables.
Key to notes:

| | |
|---|---|
| c | corrected emission |
| x | at 7.6 K |
| RT | room temperature |
| l | absorbance at low temperature |

*Solvent*

| | |
|---|---|
| ac | acetone |
| alc | 4 : 1 ethanol : methanol |
| aq | aqueous solution |
| aq-eth | 1 : 1 water : ethanol |
| DMF | N,N-dimethylformamide |
| DMSO | dimethylsulfoxide |
| EMPA | 5 : 5 : 2 diethyl ether : 3-methylpentane : ethanol |
| eth | ethanol |
| glass | porous Vycor glass |
| gly | 2 : 1 ethyleneglycol : water |
| MeCl | dichloromethane |
| MeOH | methanol |
| MF | N-methylformamide |
| nitrile | acetonitrile or other nitrile |
| pc | propylene carbonate |
| S | solid |

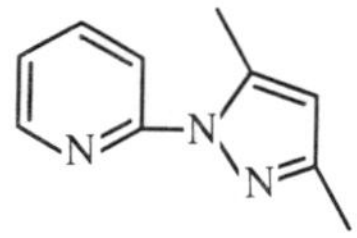
**1**

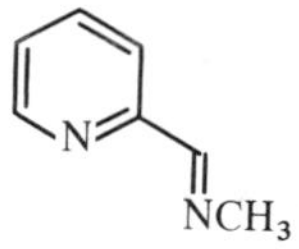
**2**

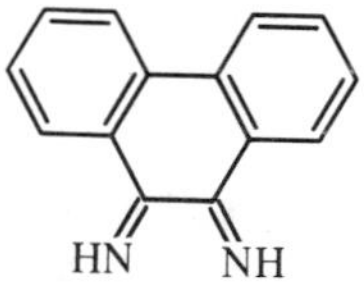
**3**

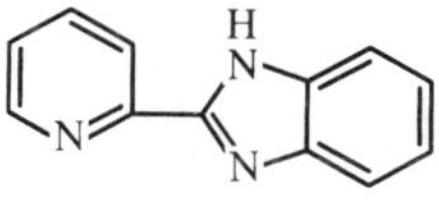
**4**

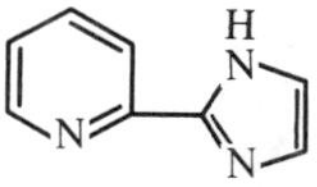
**5**

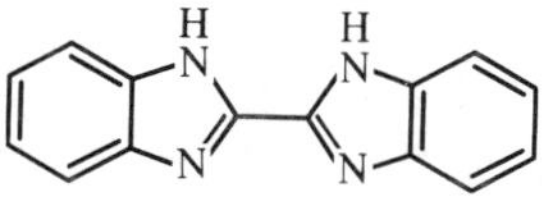
**6**

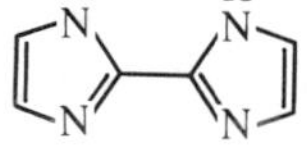
**7**

**8**

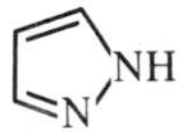
**9**

**10**

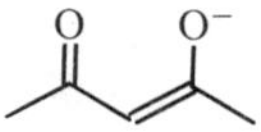
**a**

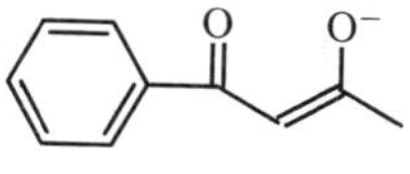
**b**

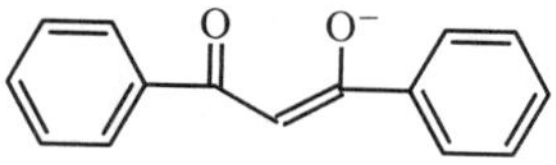
**c**

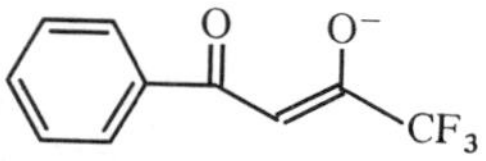
**d**

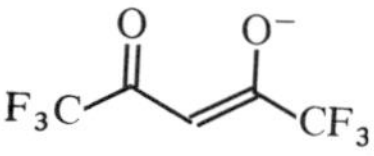
**e**

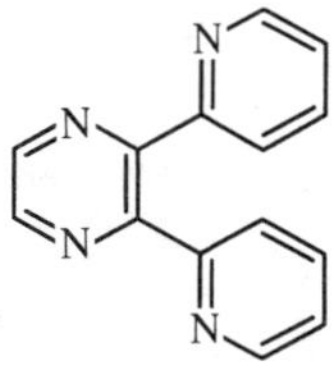
**11**

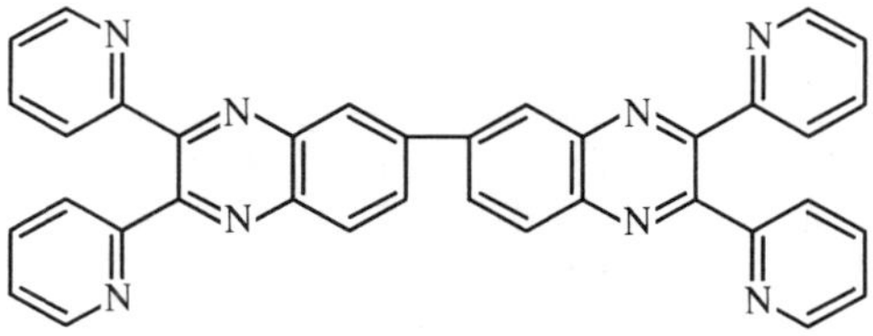
**12**

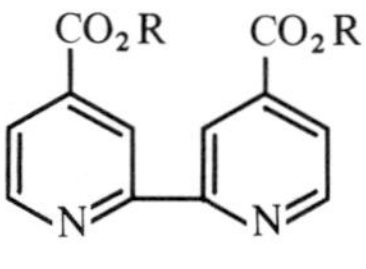
**13**

R = (a) $CH(CH_3)_2$
(b) cyclohexyl
(c) $CH_2$– 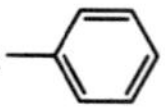
(d) β-naphthyl
(e) decalin
(f) 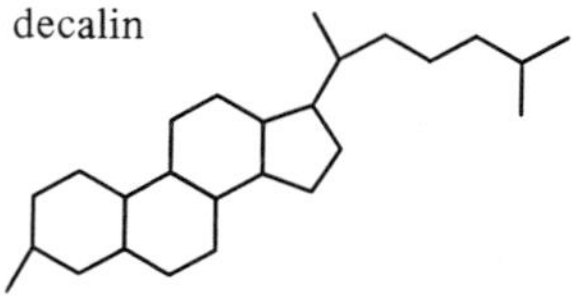

**Fig. 1**

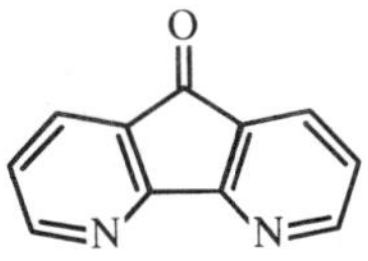
**14**

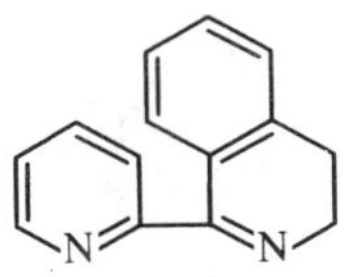
**15**

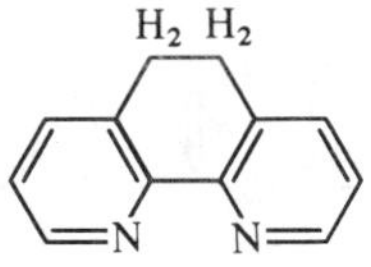
**16**

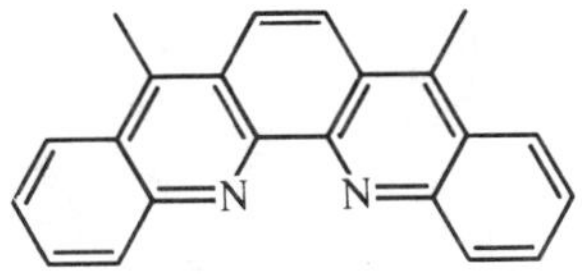
**17**

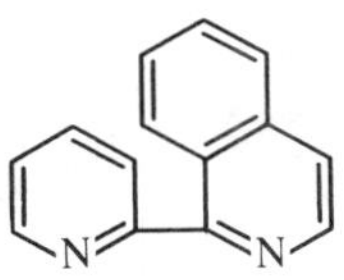
**18**

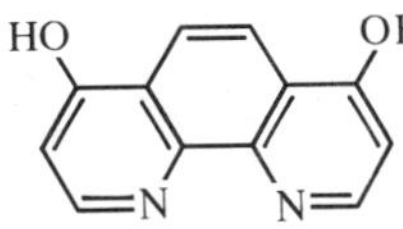
**H19**

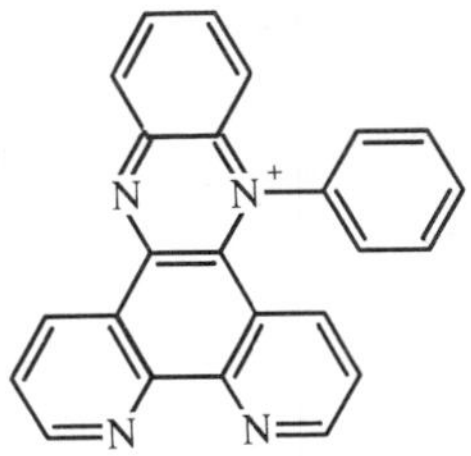
**20**

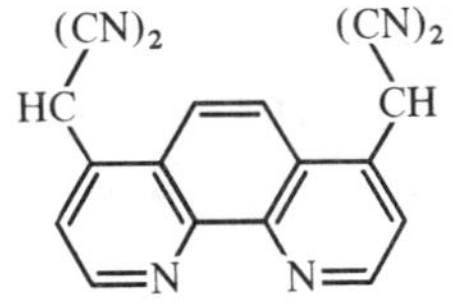
**H21**

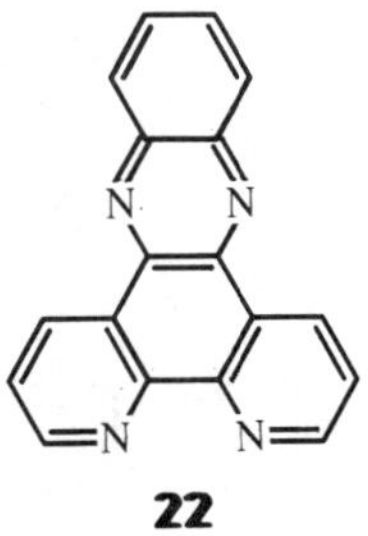
**22**

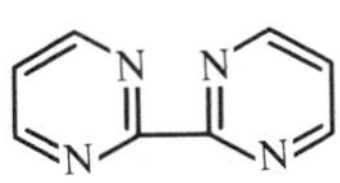
**23**

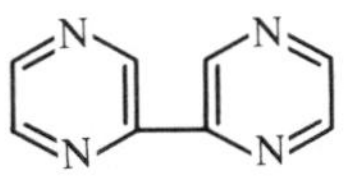
**24**

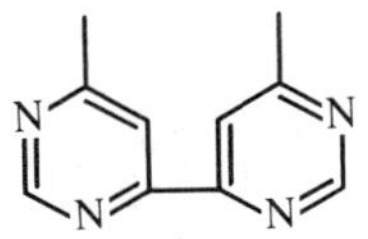
**25**

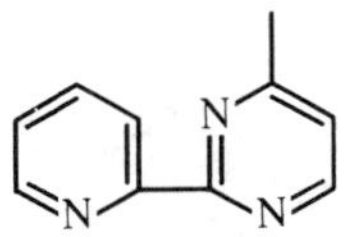
**26**

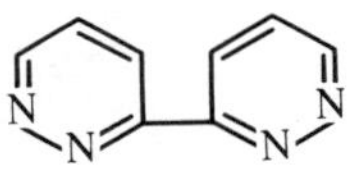
**27**

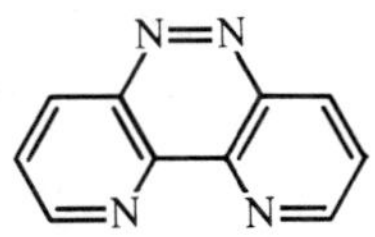
**28**

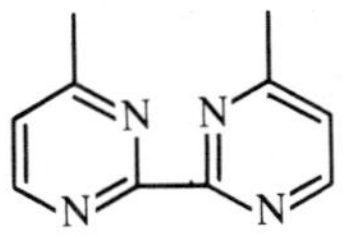
**29**

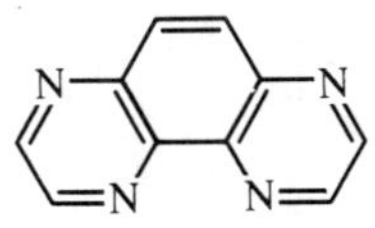
**30**

**Fig. 1**
(continued)

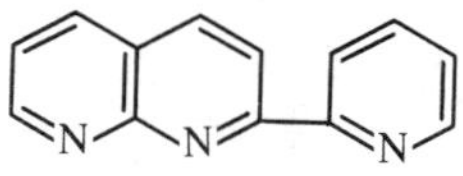

**31**

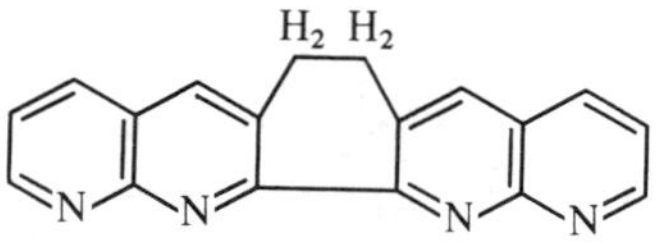

**32**

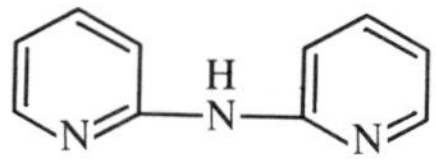

**33**

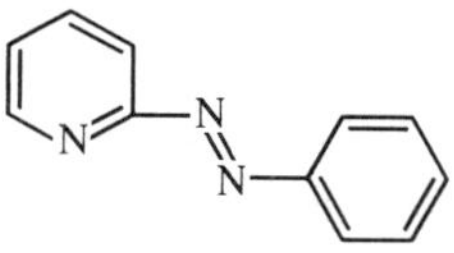

**34**

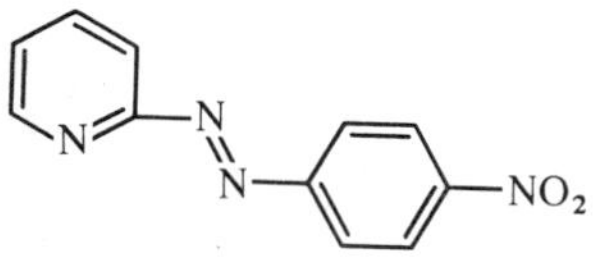

**35**

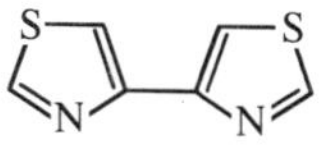

**36**

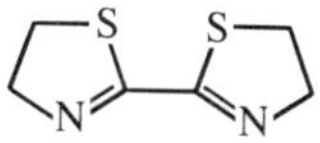

**37**

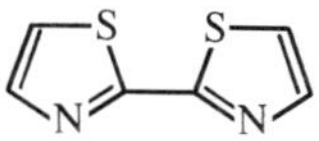

**38**

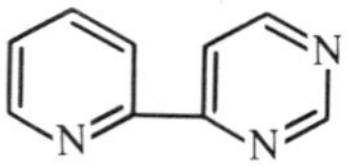

**39**

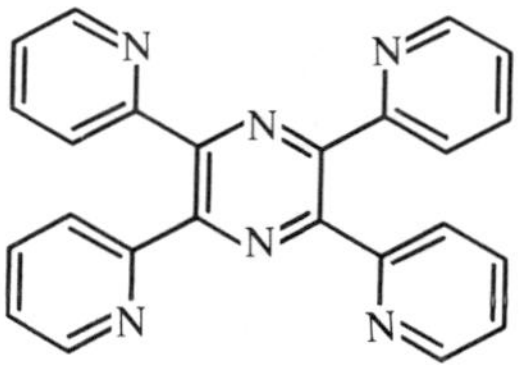

**40**

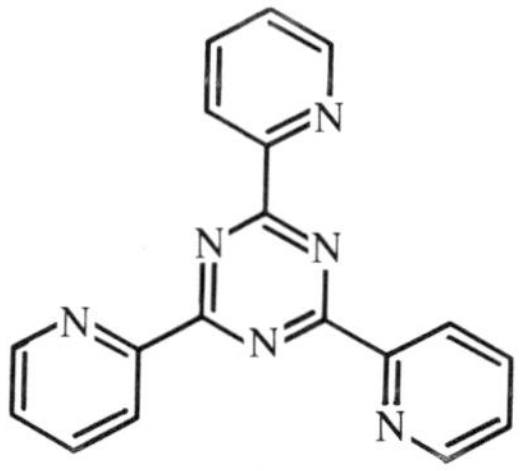

**41**

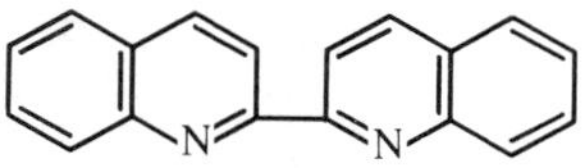

**42**

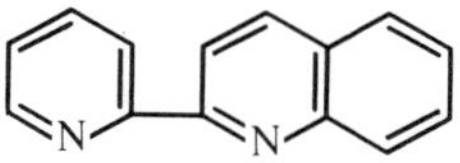

**43**

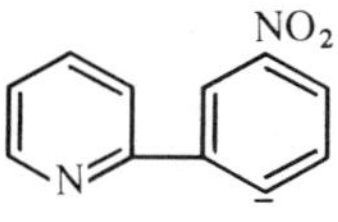

**44**

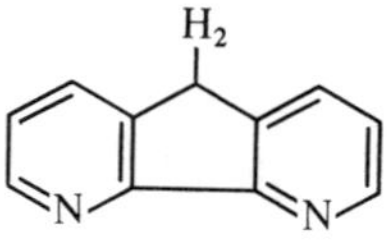

**45**

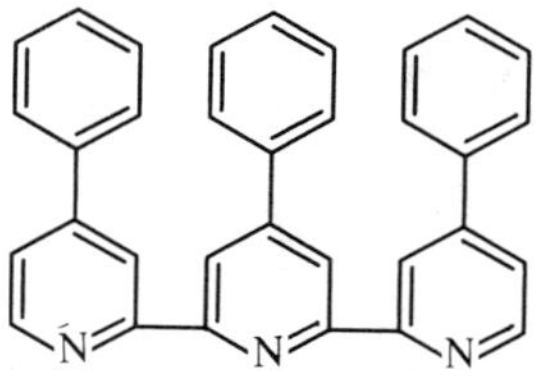

**46**

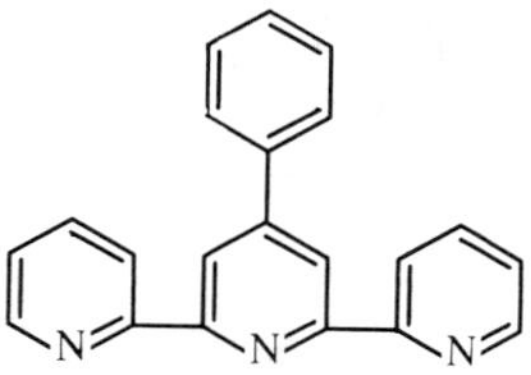

**47**

**Fig. 1** (continued)

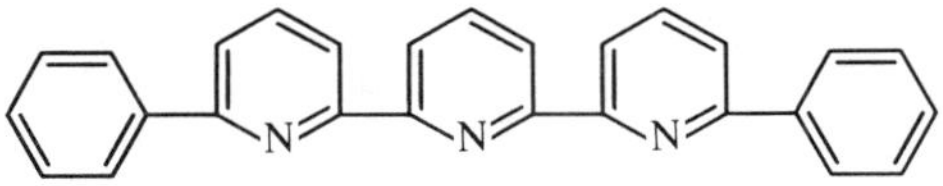


**48**

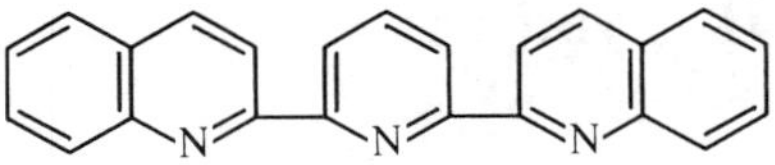


**49**

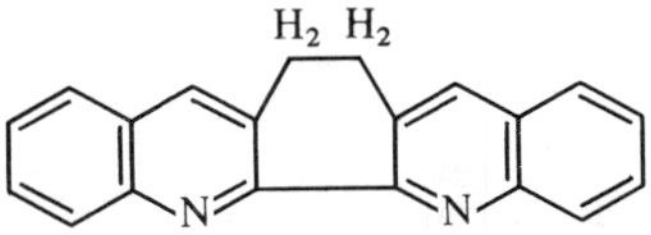


**50**

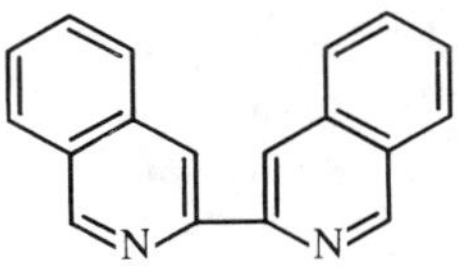


**51**

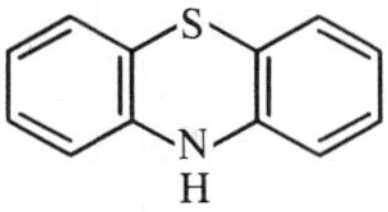


**52**

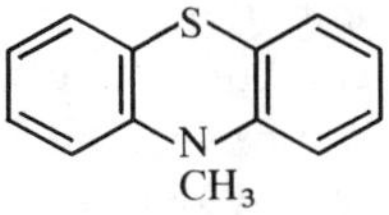


**53**

$Cl_3Pt(C_2H_4)^-$

**Pt1**

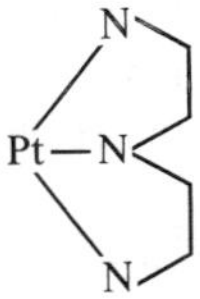


**Pt2**

**Fig. 1** (continued)

# I. Introduction

The purpose of this article is twofold: to examine how a particular group of ruthenium(II) complexes may be synthesized and to consider emission properties of these compounds as a function of molecular structure. It is our hope that ligand features important to efficient emission in this class of compounds will be somewhat clarified, and that guidelines for the synthesis of specific complexes will be laid out. While this review is not meant to be exhaustive we have attempted to include the major developments in synthesis and emission studies up through the end of 1985.

In the first section, we have summarized the principle synthetic methods which have been used, and have included a number of under utilized precursors which may be developed into generally useful materials. In the second section we consider models which have been developed for the emitting state, and how molecular structure appears to influence emission behavior. In short, what ligand features lead to an efficient emitter. This section may raise more questions than answers; however, taken together this review should facilitate determining those molecular features which might be desirable for a particular photochemical result and selecting the optimum preparative route for obtaining the desired compounds.

# II. Synthesis

## A. General Synthetic Considerations

Synthesis can be considered to consist of three separate components:
1) the reaction forming the complex
2) isolation
3) purification

While the latter two are not within the scope of this review, it will be helpful to briefly treat some general aspects of each of these three components here.

In conducting the reaction, the ligand and ruthenium source material should be at least somewhat soluble in the reaction solvent. The properties of the ligand can be important in solvent choice; solubility and stability under the employed temperature conditions may be deciding factors. An easily oxidized ligand may preclude the use of any source material containing ruthenium with an oxidation state greater than two, and readily reduced ligands may preclude the use of added reductants. Extensive tar formation is an indicator of unsuitable reaction conditions.

Isolation, once the reaction is over, usually utilizes solubility properties. Uncharged complexes are frequently insoluble in water. Rarely are coordination compounds soluble in diethylether or hydrocarbons. Thus, the addition of water, ether, or hydrocarbons (if miscible with the reaction medium) precipitates the product and many other components of the solution.

When the product is a cation a common practice is to add water (in which chloride salts are generally soluble) followed by a large anion. The most commonly used of these today are perchlorate and fluorophosphate, both of which have disadvantages. While perchlorates generally crystallize nicely, they have a troubling tendency to explode. Consequently, perchlorates are usually isolated only on a milligram scale without scraping or dry grinding of the compounds. Even in solution this anion can oxidize some ruthenium(II) compounds. Fluorophosphates generally precipitate well from aqueous solutions, but the anion tends to decompose over a period of time and may etch sample bottles. Fluoroborates do not precipitate complexes as well, at least in our hands, and suffer the same drawback.

Both perchlorate and fluorophosphate salts tend to show organic solubility (e.g., acetone, acetonitrile, occasionally methylene chloride) and are thus useful for further purification and solution measurements. Other salts can generally be prepared by metathesis. Dissolving a complex fluorophosphate salt in acetone followed by addition of a soluble quaternary chloride usually precipitates the insoluble chloride as a tar or glass, soluble in water. A similar result may sometimes be achieved by stirring a fluorophosphate salt with silver nitrate in water, precipitating sparingly soluble $AgPF_6$.

Purification of complexes is most easily achieved by recrystallization. Most workers have used the mixed solvent approach. The compound is dissolved in one solvent and a second solvent, in which the complex is insoluble but which is miscible with the first, is added. Success depends on a number of factors, one being how badly the product is contaminated. Several repetitions may be necessary. While chromatography has been used in a number of instances, we generally prefer recrystallization as being less time consuming.

It is extremely useful to have available a technique for rapidly monitoring reaction progress or assessing sample purity. An ideal technique for these purposes is thin layer chromatography (TLC). The method is fast, very sensitive, readily set up, and inexpensive. Uncharged complexes can generally be developed on silica gel plates with a variety of solvents; recently we have had good success with ethylacetate.

Cationic complexes in the past were nearly impossible to develop on silica plates. Meyer's group[1] found cations could be developed on alumina plates using a benzene-acetonitrile developer. Anderson and Seddon[2] also used alumina plates with ethanol-water developer. However, the best general developers for cationic complexes, which we have presently found, are tetraethylammonium bromide in ethanol[3] or tetraethylammonium perchlorate in acetonitrile[4]. These are used with the readily available silica gel TLC plates.

## B. *Tris-Chelates, $[Ru(AB)_3]^{2+}$ and Bis-Chelates, $[Ru(ABC)_2]^{2+}$*

A number of different methods have been employed during the past 50 years for the synthesis of tris-bidentate or bis-tridentate homochelated ruthenium(II) complexes. An examination of the various methods may lead the reader to a method best suited for a particular situation.

The first preparation of $[Ru(bpy)_3]^{2+}$ by Burstall[5] involved fusion of $RuCl_3$ in bpy, followed by extraction of excess ligand and crystallization of the product from water. Morgan and Burstall employed the same method for $[Ru(trpy)_2]^{2+}$[6]. This method is wasteful of ligand; Cook et al.[7] commented on this fact and adopted other procedures. The fusion technique has not been extensively utilized.

Palmer and Piper[8] adopted a technique originally rejected by Burstall, involving long reflux times (72 h) of $RuCl_3 \cdot 3\,H_2O$ and excess ligand in 95% ethanol. High yields of pure material can be obtained. A number of workers have used the method with other ligands[7, 9–11]. Synthesis of the 2-picolylamine complex[12] reportedly required only 12 h reflux; it is possible that excess ligand served as the reducing agent giving a faster reaction. The picolylamine ligand in $[Ru(bpy)_2(AB)]^{2+}$ can be oxidized to the imine[13] leaving some question as to the exact nature of the reported compound. However, the infrared spectrum did not indicate the presence of an imine linkage.

Somewhat shorter reaction times (ca. 30 h) have been found using ethanol-water mixtures[14, 15]. This of course requires that the ligand be sufficiently soluble in the aqueous medium. While most polypyridyls are insoluble in water, sufficient solubility may exist at the boiling point. Liu et al.[16] performed a partial asymmetric synthesis of $[Ru(bpy)_3]^{2+}$ from $K_2[RuCl_5(H_2O)]$ in water containing d-tartrate. Only 2 h heating was required; the reducing agent most likely was the tartrate. Much shorter reaction times are required when $NaH_2PO_2$ is employed as a reductant. Dwyer and co-workers[17, 18] used $K_4[Ru_2Cl_{10}O]$ in weakly acidic aqueous solution containing the ligand; addition of sodium hypophosphite yielded tris-chelates after a few minutes refluxing. The same procedure has been employed with $K_2[RuCl_5(H_2O)]$[11, 19] or $RuCl_3 \cdot 3\,H_2O$[20, 21], as source materials. Broomhead and Young[22] recommended oven drying of commercial $RuCl_3 \cdot n\,H_2O$ to achieve consistent results in the synthesis of $[Ru(bpy)_3]Cl_2$ through this route. Complexes of tridentate ligands have also been obtained in this fashion[23, 24].

Lin et al.[19] dissolved less soluble ligands in DMF before adding them to aqueous $K_2[RuCl_5(H_2O)]$. Others have followed this procedure[2,7,25]; however, Anderson and Seddon[2], in their evaluation of three synthetic methods, observed that hypophosphite reduction generally leads to products with a phosphorous impurity. And the strong reductant employed could react with some ligands; Dwyer[18] felt this to be the cause of his inability to obtain a pure complex with 5-nitro-phen. We feel that unless TLC methods are used to ascertain product purity hypophosphite reduction should be considered as an unreliable procedure.

A slight variation on the above procedures uses a ruthenium-chloro-compound in a higher boiling polyalcohol; here, solvent functions as the reductant and the higher temperatures foster more rapid reactions. This technique was apparently first utilized by Dwyer and co-workers in the synthesis of an osmium(II) complex, $[Os(phen)_3]^{2+}$[26]. Later workers have used $RuCl_3 \cdot 3\,H_2O$ and ligand suspended in ethylene glycol to obtain tris-chelates after only 30 min reflux[27], while a suspension of the reactants in glycerol gave the product after 10 h at 100°[28].

On heating in glycerol $K_2[RuCl_5(H_2O)]$ slowly dissolves. The resulting solution has been used to prepare tris-chelates: 70°, 6 h[29], one day at 100°[30], and 70° for 1 h[31]. Klassen et al. were apparently the first to adopt this method to ruthenium complexes in their preparation of a bis-tridentate complex[32]; they report 2 h reflux in glycerol for reaction conditions. $K_4[Ru_2Cl_{10}O]$ has also been used[33] simply by heating a mixture in glycerol over a burner until an orange solution resulted. Since it is frequently difficult to crystallize a complex from viscous solvents, the above reactions in polyalcohols are always diluted with water before precipitation of final product on workup.

In one report $RuCl_3 \cdot nH_2O$ was refluxed in DMF with the ligand to obtain $[Ru(trpy)_2]Cl_2$, $[Ru(bpy)_3]Cl_2$ and $[Ru(phen)_3]Cl_2$[34]. Since it is now known (and will be discussed below) that use of this solvent leads to carbonyl complexes as major components we do not feel this to be a generally useful technique. However, it does have the advantage of being very easy and using the readily available source, $RuCl_3 \cdot nH_2O$.

While $[Ru(DMSO)_4Cl_2]$ generally loses only two of the DMSO groups in an inert solvent[35], Crutchley and Lever found refluxing this compound in water with bipyrazyl for 20 h led to a tris-chelate[36]. Kirsch-DeMesmaeker et al. refluxed $[Ru(DMSO)_4Cl_2]$ in 1 : 1 ethanol-water for 20 h to also obtain tris-chelates[37,38]. Methanol was found to be an effective solvent with this precursor in preparing a complex of a tridentate isoindoline, $[Ru(ABA)_2]$, by Marks et al.[39]. Presumably the same source compound and solvents could be used for the synthesis of a wide variety of complexes. This route has the advantage of requiring no reductant and presumably not oxidizing fragile ligands, but the disadvantage of a solvent in which some ligands are insoluble.

Wilkinson's group examined the use of blue solutions of ruthenium[40] in the synthesis of a variety of complexes[41]. The reagent is generated from $RuCl_3 \cdot 3\,H_2O$ in methanol under hydrogen (2 atm) and the solution used in further synthesis. Later workers have added this solution to a ligand in glycerol or ethyleneglycol[11,29,42,43], n-propanol[2] or water[44]. After distilling off the methanol the reaction mixture is heated at a higher temperature. Anderson and Seddon[2] urged workers to adopt the "blue solution" method where trace impurities may be critical in spite of the greater intricacy and lower yields involved.

A different approach to "blue solutions" has been utilized by Bottomley. $RuCl_3 \cdot 3\,H_2O$ was first "activated" (heating and evaporation of a solution in HCl, followed by ethanol-water) and treated with aqueous formic acid[45] yielding a deep blue solution. The ruthenium(III) complex, $[Ru(bpy)_2Cl_2]Cl$[46] was prepared by this route. Bottomley's reagent, which may be quite different from the Rose and Wilkinson blue solution, has yet to be applied to tris-chelate ruthenium(II) synthesis.

One further method to produce tris-chelates requires the use of a bis-chelate complex having two replaceable ligands. This will be considered later.

### 1. *General Conclusions*

For synthesizing tris-chelates (or bis-tridentate chelates) the ideal source material is $RuCl_3 \cdot 3\,H_2O$ simply because of its availability. Hypophosphite reduction in the presence of the ligand provides a fairly general, rapid synthetic technique. While there will always be questions of purity for compounds prepared by this route, it is a valuable method for the initial synthesis of a compound to determine general properties before exploring other synthetic routes.

Many workers have employed $K_2[RuCl_5(H_2O)]$ or $K_4[Ru_2Cl_{10}O]$ as source materials. A convenient procedure for preparing the former compound starts with the commercially available $RuCl_3 \cdot 3\,H_2O$[29, 47]. A synthesis of the oxo-bridged dimer, $K_4[Ru_2Cl_{10}O]$, starting from ruthenium metal has been given by Broomhead et al.[48]; however, it is not clear to us that either of these compounds possess significant advantages over $RuCl_3 \cdot 3\,H_2O$.

Because of variations in the solubilities and stabilities of ligands, a knowledge of an array of synthetic techniques is essential. Different combinations of the procedures described above may be necessary to obtain specific compounds. A number of known source materials have yet to be fully utilized and will be considered below.

## C. *$[Ru(AB)_2X_2]$ and $[Ru(AB)_2L_2]^{n+}$ Type Compounds*

### 1. *$[Ru(AB)_2Cl_2]$ Complexes*

The dichloro-bis-bidentate complexes are a valuable group of compounds in that they are potential source materials for the synthesis of further complexes. The first synthesis of $[Ru(bpy)_2Cl_2]$ and the phen analogue was reported by Dwyer et al.[49], who isolated the product after several extractions of the pyrolysis mixture of $(LH)[RuCl_4L]$. Davies and Mullins[50] obtained the dichloro-compound by a cleaner route, which involved refluxing $(bpyH)[RuCl_4(bpy)]$ in DMF; however, no yield was reported. Bryant et al.[51] aware of carboxy products in DMF, reduced $(bpyH)[RuCl_4(bpy)]$ with zinc in hydrochloric acid to obtain the complex.They prepared the 4,4′-Me-bpy compound through the same procedure. Klassen[52] also utilized this technique for the synthesis of the pyridyl-quinoline analogue. Ability to isolate the salt, $(LH)[RuCl_4L]$, is perhaps the only restriction on this becoming a general preparative route for $[Ru(AB)_2Cl_2]$.

What appears to be a more general method for preparation of the dichloro-complexes of bidentate ligands consists of refluxing $RuCl_3 \cdot 3\,H_2O$ in DMF with stoichiometric amounts of the ligand. Whitten's group[53] detailed this method in 1977; in the following year Meyer's group[1] published a similar preparation, which is a bit more convenient, since it does not involve concentration of the reaction solution. Wrighton's group[54] described a procedure, essentially the same, that year. Some generality for this method has been demonstrated by preparing dichlorobis-chelates with substituted bpy's[55], phen's[54], and bipyrazine[56]. Nakamaru[10] used a slight modification (addition of phosphate buffer) to this method.

While carbonyl compounds are produced in the DMF reaction, they should not be important contaminants in the isolated products. Methods based on the "Whitten procedure" lead to the *cis*-isomer[57]; however, a *cis-trans* mixture of $[Ru(bpy)_2Cl_2]$ can be obtained by reacting *trans*-$[Ru(bpy)_2(py)_2](ClO_4)_2$ with hydrochloric acid[58].

Other, less important, synthetic routes to $[Ru(bpy)_2Cl_2]$ are the thermal decomposition of $[Ru(bpy)_3]Cl_2$ adsorbed on quartz[59], reaction of an oxalato complex with hydrochloric acid[60], and reaction of a hydrogen-reduced solution of $RuCl_3 \cdot 3\,H_2O$ in dimethylacetamide with bpy[61].

In a number of instances $[Ru(AB)_2Cl_2]$ complexes have been prepared through routes which customarily lead to tris-chelates. Arylazopyridines heated in ethanol with $RuCl_3 \cdot 3\,H_2O$ rapidly form the dichloro-compounds[4, 62, 63]. The strong pi-withdrawing properties of these ligands makes substitution of the remaining chlorides much slower in alcohol. Other ligands, and the technique employed, to yield dichloro-bis-chelates are bpy's disubstituted with strongly electron withdrawing groups through the Palmer and Piper method[64]; 1-(2′-pyridyl)-2-(diphenylphosphino)ethane with $K_2[RuCl_5(H_2O)]$ in aqueous ethanol[65]; 2-picolyldimethylarsine using the hypophosphite technique[66]; and 2-picoline thioamide in aqueous ethanol containing hydrazine sulfate[67]. The same explanation for production of the bis-chelates most likely holds in these cases as with the arylazopyridines.

Steric difficulties arise in synthesizing tris-chelates of large polypyridine ligands. When 2,2′-biquinolyl (ligand **42**) is refluxed in hydrochloric acid with $RuCl_3$, $[Ru(bq)_2Cl_2]$ is the product[52]. This further emphasizes the fact that a general synthesis for a complex type must take into consideration a number of factors, including the nature of the ligand.

Two compounds, related to $[Ru(bpy)_2Cl_2]$, should be considered here as potential precursors to $Ru(bpy)_2$ containing complexes. Meyer's group[68] has isolated the chloro-bridged dimer, $[Ru(bpy)_2Cl]_2(PF_6)_2$ and propose the compound as an intermediate for further syntheses. Also the ruthenium(III) complex, $[Ru(bpy)_2Cl_2]Cl$, which is readily synthesized[46] has been used as a starting compound[69] and may have general utility. Other bis-chelated compounds may be available by the routes described for these complexes.

## 2. *Other Intermediates, $[Ru(AB)_2Y_2]^{n+}$*

While the dichloro-bis-chelates summarized above are useful intermediates there are many situations in which they are of low reactivity. Many are insoluble in water; many substitutions proceed poorly in nonionizing solvents, particularly those which solvate $Cl^-$ poorly. Consequently, it is desirable to have more reactive synthetic intermediates available.

A convenient intermediate appears to be the carbonato-complex, $[Ru(bpy)_2CO_3]$. Straightforward syntheses have been reported by Meyer's group[68] and by Tanaka et al.[70] for both bpy and phen compounds, starting from the dichloro-complexes. The compounds show some solubility in organic solvents; as anticipated acids remove the carbonato group yielding a solvent complex. For example, a suspension in acetone reacts with fluorophosphoric acid to yield a soluble, reactive acetone complex[71)].

A similar complex has been prepared with bipyrazine coligands[56)] by heating the dichloro-compound with potassium carbonate in water-ethanol. This complex is insoluble in organic media and its reactivity has not been reported; it should be an excellent precursor to mixed chelates through the aquo ion.

In acidic aqueous solution the carbonato-compound $[Ru(bpy)_2CO_3]$ yields the diaquo complex which can be isolated[72)]. Davies and Mullins had obtained this aquo complex by reaction of the dichloro-compound with silver ion in aqueous solution[50)]. The interesting point in this latter work is that aquation is very rapid; thus, if water is an acceptable solvent the dichloro-compound in hot aqueous solution may be the best reagent.

Diaquo-complexes have been shown to be useful precursors with other bis-bidentate systems. We generated $[Ru(Azpy)_2(H_2O)_2]^{2+}$ from the nitrite in acid[3)] and used the solution for further synthesis. Chakravorty and co-workers have isolated the diaquo complex $[Ru(Azpy)_2(H_2O)_2](ClO_4)_2$[73)], which was prepared by treatment of the corresponding chloro-compound with silver perchlorate. This is a versatile reagent because of good solubility properties but it *can* explode. Even with very strong ruthenium(II)-chloride bonds due to the presence of good pi-withdrawing coligands, silver ion will generally remove the halide[4)].

In nonaqueous solvents silver-ion assisted reactions can be quite useful. Sullivan et al.[74)] removed chloride from $[Ru(bpy)_2Cl_2]$ in acetone with silver perchlorate; the intermediate then reacts rapidly with added ligands, olefins in this example. An apparently useful reagent is generated from $[Ru(bpy)_2Cl_2]$ in dimethoxyethane using silver fluorophosphate[75)] with the intermediate solvent complex showing a high degree of reactivity.

A compound which might be more often utilized for further syntheses is the oxalato complex, $[Ru(bpy)_2ox]$. This can be prepared from $K_2[RuCl_5(H_2O)]$[59)], (or from $K_2[RuCl_6]$[76)]) potassium oxalate and bpy. The compound is insoluble in water but dissolves in methanol and ethanol; it has been used in the synthesis of $[Ru(bpy)_2(CN)_2]$[76)].

## a) Nitrite Complexes

In cases where an aquo or carbonato complex cannot be isolated, synthetically useful nitro-complexes may be. Compounds of the type, $[Ru(AB)_2(NO_2)_2]$, have been prepared from the dichloro-complex (bpy or phen[77)], bipyrazine[56)], or phenylazopyridine[3)]). Meyer's group[78)] has demonstrated the synthetic aspects of conversion of $[Ru(bpy)_2(NO_2)X]$ into a nitrosyl followed by destruction of $NO^+$ with azide ion. When the resulting solvent complex contains acetone, a labile material is obtained. Starting with $[Ru(bpy)_2(NO_2)_2]$ both nitrite ligands may be replaced stepwise[79)]. With the stronger pi-acid ligand, phenylazopyridine, the nitrosyl is not stable giving the diaquo-complex, which is readily substituted[3)].

## b) $[Ru(AB)_2(CO)X]^+$ and Related Complexes

The carbonyl complexes, e.g. $[Ru(bpy)_2(CO)Cl]^+$, deserve separate mention, not because of the volume of work (relatively little) but because of their synthetic potential.

Kelly's group[80] reported that the "Whitten preparation" of $[Ru(bpy)_2Cl_2]$[53] produced 30–40% of $[Ru(bpy)_2(CO)Cl]^+$ which they isolated as the perchlorate salt; the carbonyl results from decarbonylation of the DMF solvent. Cole-Hamilton[81] obtained $[Ru(bpy)_2(CO)Cl]Cl$ by evaporation of the "Whitten preparation". In $[Ru(bpy)_2(CO)Cl]^+$ only the chloride ion is thermally substitutable, while only the carbonyl is replaced on photolysis[80]. Meyer's group[82] recently reported 93% yields of $[Ru(bpy)_2(CO)Cl](PF_6)$ through a rather simple procedure: the reflux of $[Ru(bpy)_2Cl_2]$ in methoxyethanol under CO. The same reaction in methanol requires longer reflux times and gives a lower yield[83]. In refluxing formic acid the carbonato complex[82] led to the isolation of $[Ru(bpy)_2(CO)(HCO_2)](PF_6)$, also in high yield. This is readily converted to a hydrido-compound, $[Ru(bpy)_2(CO)(H)](PF_6)$, the same complex isolated earlier by Kelly and Vos[84].

Later Kelly's group prepared $[Ru(bpy)_2(CO)_2]^{2+}$[85] by a stepwise synthesis involving the addition of bpy to a red carbonyl solution of $RuCl_3$[86]. Again, the carbonyl groups are photolabile. Black et al.[87] had prepared this complex by a less direct route; the Kelly procedure appears to be straightforward and has been developed to yield a number of compounds[88]. Cole-Hamilton's group[83] reported similar syntheses; this last paper nicely summarizes earlier work in the area.

### 3. *$[Ru(AB)_2(L)_2]^{n+}$ Complexes; General Conclusions*

Precursors described in the preceding section have been used to prepare the majority of complexes of the $[Ru(AB)_2(L)_2]^{n+}$ type. The desirability of first isolating a compound containing the $Ru(AB)_2$ moiety must be emphasized: while $[Ru(bpy)_3]^{2+}$ is formed stepwise[89], the *in situ* generation of $Ru(bpy)_2$ never gives assurance of the absence of the tris-chelate. To date, the single most general precursor in this class is $[Ru(AB)_2Cl_2]$, but there is no guaranteed synthesis for $[Ru(AB)_2Cl_2]$ compounds with a given ligand. Our own preference is to try several routes; however, the most widely applicable is probably the DMF procedure[1].

Solvent complexes, $[Ru(AB)_2(solv)_2]^{2+}$, which may be generated in several fashions can also be effective precursors. Perhaps the most general route to these is through the use of silver ion to remove chloride from $[Ru(bpy)_2Cl_2]$ in acetone[1]. Meyer's paper noted the disadvantage of coordinated acetone, which is capable of undergoing nucleophilic attack[75], and developed a similar procedure in dimethoxyethane. It should be noted that while the dichloro-complex is essentially insoluble in both of these solvents, the silver-ion reaction proceeds smoothly. Although neither route has been developed for other chelates, both should be widely applicable.

Silver ion has also been utilized for chloride removal in aqueous media[4,73]. Again, ligand insolubility can be a deciding factor in use of this solvent.

## D. *Monochelates, $[Ru(AB)L_4]^{n+}$ and $[Ru(ABC)L_3]^{n+}$*

An important group of compounds from the standpoint of further syntheses, particularly mixed chelates, are the monochelates. In 1938 Morgan and Burstall[90] reported what were probably the first members of this group to be isolated, $[Ru(NO)bpyCl_3]$ and $[Ru(NO)(trpy)Cl_2]Cl$; these compounds do not appear to have been extensively utilized for further synthesis. Dwyer et al.[91] obtained the bpy and phen complexes, $[Ru(AB)Cl_4]$ and $[Ru(AB)Cl_4]^-$, the latter ion being isolated as potassium, ammonium, or protonated ligand salts. Klassen[52] utilized a similar procedure to obtain (pqH)$[Ru(pq)Cl_4]$. While the ligand salt species leads to $[Ru(AB)_2Cl_2]$ it does not appear to have much use as a precursor to mixed chelates containing only one of the original chelating ligands. James and McMillan[92] reported an improved one-step route to K$[Ru(bpy)Cl_4]$ (or the phen analogue) which involved initially refluxing $RuCl_3 \cdot 3\ H_2O$ in methanol under hydrogen. However, the easiest route to a monochelated bpy compound is one we reported in 1977[93]. This was initially called $[Ru(bpy)Cl_4]$ but in fact is $[Ru(bpy)Cl_3]_n$; it is prepared in one step and reliably leads to further complexes containing one bpy.

Ford and co-workers have prepared $[Ru(NH_3)_4(AB)]^{2+}$ through the diaquotetraammine[94] (AB = bpy, picolylamine, or picolylimine). This procedure was also utilized by Ruminski and Petersen[95] for the bipyrimidine chelate, $[Ru(NH_3)_4(bpym)]^{2+}$, and the route should have considerable generality. However, the ammine complexes may show poor non-aqueous solubility properties hindering further reactions.

Meyer's group[96] has reported a very simple synthesis of $[Ru(trpy)Cl_3]$. Unfortunately, broad generality has not been demonstrated for any of these methods and no prescription can be given at this time for a synthesis of monochelated complexes.

Evans et al.[35] reported a facile synthesis of the dimethylsulfoxide complex, $[Ru(DMSO)_4Cl_2]$, and claimed only two DMSO ligands were substituted in inert solvents. $[Ru(DMSO)_2(bpy)Cl_2]$ and the phen analogue were isolated after reaction in chloroform. Two 8-substituted quinolines were observed to yield $[Ru(DMSO)_2Cl_2(AB)]$ with toluene as solvent[97]. However, we have obtained $[Ru(AB)_2Cl_2]$ from the DMSO complex in both acetone[62] and chloroform (AB = ligand **34**). The DMSO complex does not appear to offer a reliable route to any one group of compounds, but is certainly synthetically useful.

Complexes of the chelating diolefins, norbornadiene and 1,5-cyclooctadiene, have yielded mono-bpy type compounds. The oligomer, $[Ru(nbd)Cl_2]_n$[98], with bpy in acetone gives $[Ru(nbd)(bpy)Cl_2]$[74] which yields $[Ru(bpy)_3]^{2+}$ with excess ligand in DMF. Schrock et al.[99] have investigated reactions of $[Ru(nbd)(allyl)_2]$ giving $[Ru(nbd)(AB)(CH_3CN)_2]^{2+}$ and $[Ru(AB)(CH_3CN)_4]^{2+}$ (AB being bpy, phen, or a chelating diarsine) while Robertson et al.[100] have prepared $[Ru(C_6H_6)Cl(phen)]^+$. None of these methods have been developed into general procedures for monochelated complexes but the $[Ru(nbd)(allyl)_2]$ route seems to hold promise. Removal of the allyl group requires electrophilic attack allowing its stepwise displacement and thus some assurance of substitution at specific coordination sites.

As with synthesis of the bis-bidentate chelates, carbonyl containing monochelated complexes show some promise for further syntheses. The first compound of this type, $[Ru(CO)_2I_2(bpy)]$, was prepared by Irving in 1956[101] starting from the iodo-carbonyl polymer. Wilkinson's group[102] utilized Chatt's "red carbonyl" solution in ethanol[86] to prepare $[Ru(CO)_2(AB)Cl_2]$ (using bpy and phen); no further substitution with excess ligand was possible. Kelly et al.[85] obtained the same compound but found that the addition of water yielded a different isomer which was more reactive; this isomer would add a second molecule of a bidentate ligand.

Cole-Hamilton's group[83] prepared $[Ru(CO)Cl_2(trpy)]$ by refluxing $RuCl_3 \cdot xH_2O$ in DMF with the tridentate ligand for 24 h. But Deacon et al.[103] isolated isomers of $[Ru(CO)_2X_2(trpy)]$ containing *bidentate* trpy after briefly heating the halocarbonyl in methanol with the ligand. Decarbonylation with trimethylamine oxide gave [Ru(CO)$X_2$ (trpy)] containing tridentate trpy.

Hiraki et al.[104] have prepared a yellow carbonyl solution in methoxyethanol. Addition of 2-phenylpyridine and related ligands led to the formation of orthometallated monochelates, [Ru$(CO)_2$Cl(AB)$]_2$. Addition of further ligands then yielded mononuclear complexes.

In the forgoing syntheses the carbonyl group was present in the precursor or introduced during the synthesis (DMF preparations). Meyer's group[96] has added CO to both *cis*- and *trans*-[Ru(trpy)(P$(ph)_3$)$Cl_2$] to obtain *cis* and *trans* [Ru(trpy)(CO)$Cl_2$], the phosphine being replaced with retention of stereochemistry.

It appears to us that the most simple, and best developed reagent for the carbonyl-chelates is the "red carbonyl" solution. A "convenient method" has been reported for preparing such solutions[105], but since this method requires the preparation of coconut charcoal, it may be preferable to employ the original Chatt procedure.

Other monochelated complexes are known. If the precursor contains moderately good ligands for ruthenium, it is reasonable to expect limited substitution by the incoming chelating group, which makes further substitution difficult. Gilbert and Wilkinson[106] have demonstrated the synthetic utility of the five-coordinate triphenylphosphine complex, [Ru$Cl_2$$(PR_3)_3$]. Switkes et al.[107] have shown a variety of reactions to occur with the triphenylphosphine or -arsine complexes of ruthenium(III), [Ru$Cl_3$$(MR_3)_2$(Solv)]. Reaction with bpy in methylene chloride yields [Ru$Cl_3$$(MR_3)$bpy] while in methanol [Ru$Cl_2$ $(bpy)_2]^+$ is claimed. Again, this has not been generalized.

Another example of a mono-bpy compound was prepared by Gilbert et al.[41] from the "blue solution". A benzonitrile complex was isolated which, with bpy in methanol, yielded [Ru$Cl_2$(bpy)$(RCN)_2$]. No further substitutions were reported on this compound, but it might be developed into a useful synthetic tool.

## E. Mixed Chelates

For a variety of purposes it is useful to have complexes containing more than one kind of chelating ligand. A great many complexes of the type [Ru$(AB)_2$(CD)$]^{n+}$ have been reported, with the majority containing a bpy or phen derivative as one (or two) of the chelate groups. To our knowledge, few complexes have been reported containing three different bidentate ligands, although routes can be devised for such syntheses[115].

The synthesis of mixed chelates requires a precursor such as [Ru (bpy)$Cl_3]_n$ or [Ru$(bpy)_2$$Cl_2$], which were described earlier. There are fewer examples of monochelated compounds than bis-chelated. A general synthetic scheme requires addition of a bi- or tridentate ligand to the precursor. For a complex of the type, [Ru(AB) (CD)(EF)$]^{n+}$, the precursor should contain two final leaving groups which are more difficult to substitute on the last step.

At each step of the reaction the product should be isolated; otherwise the final result will surely be a mixture of compounds. As an example of this type of procedure we consider an early paper by Dwyer et al.[91]. [Ru(bpy)$Cl_4$] with pyridine in aqueous ethanol led to [Ru(bpy)$(py)_4$]$(ClO_4)_2$, which could be converted into several [Ru(bpy) $(py)_2$(AB)$]^{n+}$ compounds. Later we showed that compounds of this type, on long reflux with more bidentate ligand, yielded complexes in which the remaining pyridines were replaced[93].

The most general route to complexes of the type $[Ru(AB)_2(CD)]$ involves reactions of $[Ru(AB)_2Cl_2]$. To substitute chlorides in this type of compound requires a polar solvent, which will solvate the leaving anions. Solvents which have been successfully used include 1 : 1 methanol-water[13, 108], ethanol-water[56], water[109], acetone-water[110], and methanol[111]. There are, of course, many ligands which will not dissolve in these combinations and a different procedure is required.

More difficult situations may make use of the dimethoxyethane solvate[112]. Occasionally chlorides in $[Ru(AB)_2Cl_2]$ are not readily substituted even in alcohol-water solvents. Silver ion is then quite often useful[4] to yield solvated precursors, which have been considered earlier.

A number of reports indicate the potential usefulness of carbonyl complexes in the synthesis of mixed chelates, and significant progress has been made recently in this area by Black and co-workers. Utilizing the known reaction of trimethylamine N-oxide with coordinated carbonyls[113] they were able to convert $[Ru(CO)_2X_2(AB)]$ into $[Ru(CO)X_2(AB)(py)]$[114] (AB being bpy or phen). Later, they demonstrated that even the second carbonyl may be removed by this route in methoxyethanol in the presence of ligands to give $[Ru(AB)(CD)_2]^{2+}$[115]. It is also possible to replace only the chlorides in $[Ru(CO)_2Cl_2(AB)]$[116] by a single step procedure to yield the intermediate, $[Ru(CO)_2(tf)_2(AB)]$[117]. The triflate anion can be readily substituted yielding $[Ru(CO)_2(AB)(CD)]^{2+}$, where the bidentate ligands are bpy, phen, or their derivatives.

Kelly et al. had reported a much easier route in observing that the yellow isomer of $[Ru(CO)_2(bpy)(Cl)_2]$ would add a second bpy in refluxing water-ethanol[85]. They found the carbonyls could then be substituted by solvent (methanol, water, or acetonitrile) photochemically. However, it should be noted that $[Ru(bpy)_2(H_2O)_2]^{2+}$ photochemically isomerizes to the *trans*-form[72], which is slow to add a third bidentate ligand, requiring isomerization back to the *cis*-configuration. All of these reactions need more investigation to establish general routes to mixed chelates.

## F. Potential Precursors

A number of compounds have been proposed as useful intermediates for the synthesis of ruthenium(II) complexes. Singleton and co-workers have investigated $[Ru(diene)(N_2H_4)_4]^{2+}$ in this respect. Reaction of the cyclooctadiene compound with bpy in acetone yields $[Ru(cod)(bpy)_2]^{2+}$[118]. With some ligands complete substitution resulted[119]. They later claimed[120] $[Ru(H)(cod)(NH_2NMe_2)_3]^+$ to be the most versatile precursor known at that time.

Ludi's group has isolated the aquo ion $[Ru(H_2O)_6]^{2+}$ as a p-toluenesulfonate salt[121, 122] and shown it to have use as a synthetic reagent. The aquo groups are very labile[123], and the compound displays good solubility in organic solvents; several phosphine complexes have been prepared by reactions of this precursor in tetrahydrofuran[124]. A disadvantage is the somewhat involved synthetic method; also, strongly basic ligands decompose the compound. Bernhard and Sargeson[125], recognizing this drawback, converted the aquo ion into $[Ru(DMF)_6]^{2+}$ for further synthesis. Neither the aquo nor DMF complexes have yet been extensively utilized.

The triflate anion is recognized as being an extremely good leaving group[126]. In situations where chloro-ligands are not readily displaced, reaction with anhydrous Htf can displace the chlorides producing a more reactive intermediate. This technique will certainly be utilized more in the future; however, a disadvantage is the extremely corrosive nature of the acid. We have found pierced teflon lined silicone rubber septa to be attacked by it.

Phosphine and arsine complexes have been used as precursors in a few cases. Ruiz-Ramirez and Stephenson found $[RuX_3(L)_2(MeOH)]$ ($L = P(ph)_3$ or $As(ph)_3$) to react with bpy or phen in methylene chloride to yield $[RuX_3(L)(AB)]$, while in methanol $[RuX(L)(AB)_2]^+$ is produced[127]. Poddar and Agarwala observed $[RuL_2Cl_2]_2$ to react with pyridine to yield a number of products, including $[Rupy_4Cl_2]$[128]. While complete displacement of the phosphine or arsine can occur, reactions with these kinds of precursors generally result in one or more of the original ligands being retained in the product, and we feel these compounds to be of limited synthetic utility.

One final family of potential precursors has been reported by Newton and Searles[129], who found nitrile complexes of the type, $[RuX_2(RCN)_4]$, to be soluble in a number of solvents. The nitriles can be displaced in boiling methoxyethanol, and the benzonitrile complex was used by tom Dieck et al.[130] in boiling diglyme to prepare isomeric diazadiene complexes $[Ru(AB)_2Cl_2]$. This may represent a useful alternative route to the DMF synthesis.

## G. Photochemical Syntheses

Photochemical methods have occasionally been employed in the synthesis of ruthenium(II) complexes. While the same complexes could generally be prepared by alternate routes, the lower reaction temperatures required can be advantageous. Moreover, in some cases, the photolytic method can offer a selectivity not otherwise available.

While the lowest excited state in ruthenium(II) complexes generally involves a MLCT, a d state is nearby (this aspect will be considered later). Hoggard and Porter[131] studied the photolysis of $[Ru(bpy)_3](SCN)_2$ in DMF and ethanol; $[Ru(bpy)_2(NCS)_2]$ was cleanly isolated from ethanol in this fashion. Wallace and Hoggard[132, 133] photolyzed several $[Ru(bpy)_3]^{2+}$ salts preparing the di-anated compounds. Jones and Cole-Hamilton have developed a moderate yield (38%) photochemical synthesis of $[Ru(bpy)_2Cl_2]$ from $[Ru(bpy)_3]Cl_2$[134]; the authors did not consider product stereochemistry.

Polyatomic anions can yield complications. Broomhead and Ileperuma[135] found $[Ru(bpy)_3](NCSe)_2$ in methylene chloride yielded $[Ru(bpy)_2(SeCN)_2]$, while in ethanol the product probably was the cyanate. Wallace and Hoggard[132] had claimed the azido-complex $[Ru(bpy)_2(N_3)_2]$ to be the photolysis product in ethanol.

Crutchley and Lever photolyzed the bipyrazyl chelate $[Ru(AB)_3]^{2+}$ to obtain $[Ru(AB)_2Cl_2]$ in 44% yield[36]. They later synthesized this compound through the DMF route[56]. But a pertinent point is illustrated here: when the tris-chelate is on hand, photolytic routes may yield enough product to determine if further synthetic work is justified, and the compound is more easily recognized while developing the synthesis.

Steel et al.[21] photolyzed a pyridyl-pyrazole chelate in acetonitrile to obtain $[Ru(AB)_2(solv)_2]^{2+}$, which was used in further syntheses.

A number of workers have photolyzed $[Ru(AB)_2XY]$ complexes as a route to other compounds. Bosnich and Dwyer[136] found $[Ru(phen)_2(py)_2]^{2+}$ to be converted to

$[Ru(phen)_2Cl_2]$ rapidly in bright sunlight while the thermal reaction proceeded much more slowly. Meyer's group described photochemical syntheses of $[Ru(bpy)_2X_2]$ and $[Ru(bpy)_2LX]^+$ starting with $[Ru(bpy)_2(L)_2]^{2+}$ salts[137] (L = py or 4-t-butyl-py). They also investigated the *cis-trans*-photoisomerization of $[Ru(bpy)_2(H_2O)_2]^{2+}$, obtaining a *trans*-salt in this fashion[72].

The photolability of the carbonyl group in $[Ru(bpy)_2(CO)Cl]^+$ and $[Ru(bpy)_2(CO)_2]^{2+}$ has already been referred to, and these complexes may be developed as useful precursors.

# III. Ruthenium(II) Emission

## *A. Luminescence-Introduction*

Since the first report of $[Ru(bpy)_3]^{2+}$ emission[138] a tremendous body of literature has developed attempting to characterize the nature of this process. While several models have been developed and debated, many processes seem to be occurring and all of the models to date possess some definitely useful points. Here we shall attempt to examine the major concepts and see how the work is leading to emitter design.

Ligand design has allowed certain aspects of the $[Ru(bpy)_3]^{2+}$ problem to be examined. In many cases, instead of simplifying the overall picture, these compounds have demonstrated new phenomena, which may ultimately give rise to a very general model encompassing many aspects.

Our goal in this section is to examine the molecular features important in the design of efficient luminescent compounds. While we still know very little about the ruthenium(II) excited states and their decay kinetics, some guidelines for the molecular design of efficient emitters are beginning to develop. Kalyanasundaram, in a recent review of $[Ru(bpy)_3]^{2+}$, has nicely summarized the picture of the excited states up to 1981[139]. Here we shall briefly touch on early work in order to gain some perspective.

Following Paris and Brandt's initial report[138] there were several attempts to make an assignment of the emitting state. In 1968 Klassen and Crosby[140] assigned the luminescence as π-d, with the π state being of uncertain spin multiplicity. That same year Demas and Crosby[141] concluded the luminescence to be a spin forbidden process. In 1973 Harrigan and Crosby presented evidence that the excited state consists of a manifold of three states, separated by ca. 60 $cm^{-1}$[142]. These states were postulated as being in thermal equilibrium; an analysis of quantum yield and lifetime temperature dependence (4.2 to 77 K) gave excellent agreement with this model[143].

Crosby, Hipps and Elfring then proposed that, due to spin-orbit coupling, spin labels have no meaning in these excited states[144]. This has the effect of creating delocalized π* states in the complex ion. Crosby and Elfring developed this concept further[145] claiming strong ligand-ligand interaction to be manifested in the complexes $[Ru(bpy)_m(phen)_{3-m}]^{2+}$. By way of comparison, they demonstrated there to be only very weak interligand coupling in the analogous rhodium(III) compounds.

Thus, the Crosby model involves molecular π* states, spread out over all ligands, these states consisting of several closely spaced levels. This implies that emission properties of a molecule will be determined by the entire ligand set contributing to the π system.

If the Crosby model does not hold, then luminescence properties should be dictated by the individual chromophores in that molecule; i.e., weak ligand-ligand interaction would leave a set of individual $\pi^*$ states. Once a triplet state on one ligand is populated, transfer to states of other ligands should be poor[146)]; this raises the possibility of multiple emissions.

Evidence both for and against this coupled model has accumulated. In 1977 we reported on emission properties of a number of mixed ligand complexes of ruthenium(II)[146)], in which little coupling appeared to be present. Anderson et al. described a series of mixed complexes containing bpy and pq[147)] and concluded there to be little coupling. Later the same group examined the $[Ru(phen)_m(pq)_{3-m}]^{2+}$ series claiming the individual MLCT emissions could be detected through lifetime measurements[148)]; however, this was later shown to be incorrect[149)]. Klassen, examining a related set of complexes, concluded the CT states for each ligand are independent and emission was from the lowest emitting state only[52)]. Sullivan et al. reported multiple emissions from a mixed-ligand ruthenium(II) complex[150)] providing strong evidence against Crosby's coupled-ligand model.

Polarization measurements have been employed by a number of groups to aid in resolving this question. Hipps concluded, from emission polarization studies, that both ground and excited states cannot be of $D_3$ symmetry[151)] indicating a localized model. While DeArmond's group analyzed excitation polarization spectra[152)] and concluded emission to be from an E type state, verifying delocalization, their later studies[153, 154)] favored a localized model.

Felix et al.[155)] analyzed the absorption and fluorescence spectra of the complex doped into $[Zn(bpy)_3]SO_4$ and $[Zn(bpy)_3]Br_2$. Fluorescence polarization measurements in a rigid glass could not be interpreted on the basis of $D_3$ symmetry for the emitting state. However, they did conclude that the CT bands can be classified as spin-allowed and spin-forbidden, the intensity of the latter deriving from spin-orbit coupling. (Krol and Blasse[156)] also observed low-temperature emission from a doped $[Zn(bpy)_3]SO_4$ matrix and proposed several different ruthenium centers to explain the emission spectra from 5.3 to 105 K.) But Yersin and Gallhuber[157)], in their study of single crystal polarized emission stated there to be no reason for assuming an excited state of lower symmetry than the ground state. They proposed earlier work, supporting a localized model, was influenced by solvent interactions.

Excited state resonance raman should be a powerful technique for application to this question. Dallinger and Woodruff[158)] concluded that their raman spectra indicated the excited complex to be $[Ru(III)(bpy)_2(bpy\cdot^-)]^{2+}$ (i.e., localized). Raman spectra of the chemically generated reduced forms of the complex, $[Ru(bpy)_3]^{+2-n}$ ($n = 1$ to 3), were in agreement[159)]. In 1984 Caspar et al., using Franck-Condon analysis and resonance raman (both ground and excited state raman) calculated excited state parameters for some osmium complexes as well as $[Ru(bpy)_3]^{2+}$[160)] and concluded that the localized hypothesis is correct. However, in a recent paper Krausz[161)] noted raman results to be dependent on the medium. Solid $[Ru(bpy)_3]^{2+}$ salts, or the complex dispersed in solid matrices at low temperature, gave different results than observed in solution. These observations were interpreted as evidence that charge localization takes place rapidly in solutions from a delocalized state but is inhibited in rigid media.

Ferguson and co-workers appear to be reaching a resolution of the conflict between the two extreme points of view. A study of the complex doped into $[Zn(bpy)_3](BF_4)_2$ ($D_3$

sites) and $[Zn(bpy)_3](PF_6)_2$ ($C_2$ sites) led[162] to a new model containing localized triplets and delocalized singlets. A later model[163] claimed that a delocalized chromophore is essential. Emission must be from an E state, consistent with the conclusions of others. As in the Crosby model, three closely spaced states (but with 220 $cm^{-1}$ separation) were proposed. However, population of these states should be through three different pathways, two of which are blocked below 30 K and 6 K. Time resolved luminescence studies clearly show a rapid relaxation from one set of emitting states to another[164]; Ferguson's arguments have been recently summarized[165] and a strong case is made for both delocalized and localized luminescent states.

It has generally been assumed that the MLCT absorption is singlet-singlet followed by intersystem crossing populating the observed excited states. Crosby's assertion that the spin quantum number loses its significance[144] was generally accepted, although gradually increasing dissension can be seen in the literature. Evidence for important singlet excited states has been frequently observed; e.g., Gafni and Steinberg[166] claimed evidence for some singlet-singlet emission and Mandal and Demas[167] claimed singlet energy transfer from $[Ru(bpy)_3]^{2+}$ to dyes. Demas' group[168] felt singlet energy transfer should be interpreted as supporting the existence of spin-orbit states. Ferguson's model, containing both singlet and triplet excited states, certainly receives support from earlier observations. As early as 1976 Kirk et al.[169] claimed there to be an initial, short-lived CT state which then goes to the longer-lived CT state, quite in keeping with the later model. Recently[170, 171], Kober and Meyer developed an electronic model for the $M(bpy)_3$ complexes (M = Fe, Ru, Os). They point out that exciton theory requires a delocalized model in the initial singlet-singlet absorption process, but that vibrational equilibrium and spin-orbit coupling at the metal could rapidly populate triplet states. The conclusion was that the excited states are localized, at least in fluid solutions; four low-lying MLCT states exist, three close together and one several hundred $cm^{-1}$ higher. As further evidence for the localized model in fluid solution Kober et al.[172] studied the MLCT absorption bands in a range of solvents. Their results are consistent only with a localized model.

The overall picture at present, then, is one of several closely spaced excited states which are predominantly triplet in nature. These are populated from delocalized singlets involved in the absorption process. Some confusion exists on the importance of the singlets in excited state behavior and this could well be due to the varied conditions for the arguments. A delocalized triplet state does not seem favored by the bulk of the evidence.

A further aspect in the electronic structure of $[Ru(bpy)_3]^{2+}$ is the importance of d-d states slightly above the low-lying triplets. Demas and Adamson[173] first noted a slight photodecomposition of $[Ru(bpy)_3]^{2+}$. Van Houten and Watts[174] concluded that a d-d state lies ca. 3600 $cm^{-1}$ above the CT manifold, and thermal population of this level accounts for photolysis of $[Ru(bpy)_3]^{2+}$ at elevated temperatures. They were able to isolate a photolysis product[175]. Allsopp et al. concurred with the basic model of Van Houten and Watts but concluded that two temperature activation parameters were necessary for a satisfactory description above room temperature after studying emission lifetimes in a variety of solvents[176]. A number of workers have observed photoanation and photosolvolysis of $[Ru(bpy)_3]^{2+}$ without speculating on the nature of the labile excited state[131–134, 177], while Porter and Sparks[178] concluded that photoracemization of the complex occurs via the phosphorescing state. Most recently Durham et al.[179] agreed that photosubstitution occurs through a thermally accessible d-d excited state.

Henderson et al.[180] modified $[Ru(bpy)_3]^{2+}$ by incorporating a structurally distorted analogue of bpy, daf, into mixed complexes with bpy. This is a poorer ligator due to the $-CH_2-$ caused distortion (daf being slightly non-planar) and LF states will lie lower than in $[Ru(bpy)_3]^{2+}$. Interestingly, $[Ru(bpy)_2(daf)]^{2+}$ has a high emission quantum yield (0.56 vs. 0.33 for $[Ru(bpy)_3]^{2+}$, found under these authors conditions) at 77 K. A rapid drop in quantum yield with increasing temperature is presumably due to easier population of a d state (estimated at 3000 $cm^{-1}$ lower than $[Ru(bpy)_3]^{2+}$) from the CT manifold. In keeping with this concept, $[Ru(daf)_3]^{2+}$ is a very weak emitter at 77 K.

However, Reveco et al.[181] observed a pronounced temperature dependence for the emission quantum yield in $[Ru(bpy)_2(NPP)]^{+}$ and concluded there to be an energy separation between emitting and deactivating states. The complete absence of photosubstitution in this complex led them to propose that the deactivating state was MLCT with significant singlet character, and not ligand field. Allen et al.[182] had previously considered the possibility of this interpretation.

In the past few years several additional points have been raised in the problem of ruthenium emission. Caspar et al.[183] showed the relevance of the energy gap law to a series of nearly 30 osmium(II) complexes, all emitting at room temperature in fluid solution. Basically, the energy gap law predicts that the non-radiative rate constants will be dependent on the energies of ground and excited states, considering excited states in a series of compounds based on the same chromophore. Thus, the higher the emission energy, the lower the value of the nonradiative rate constant and the more intense the emission.

A study of emission lifetimes and quantum yields allowed the calculation of both radiative ($k_r$) and nonradiative($k_{nr}$) rate constants[183]. A plot of ln $k_{nr}$ vs. $E_{em}$ (emission energy) showed the expected linear dependence, with $k_{nr}$ being least for the maximum $E_{em}$. Two different plots were obtained depending on the chromophoric group, one for Os-bpy and another for Os-phen complexes. In the compounds studied, the $k_{nr}$ values changed ca. two orders of magnitude, while $k_r$ values spanned one order of magnitude. Thus, while $k_r$ is important, it is $k_{nr}$ which shows the greater variation with structure and hence appears to dominate the emission behavior. An important conclusion was that $k_{nr}$ would be determined largely by vibrational overlap between ground and excited states.

In the following year Caspar and Meyer[184] applied these considerations to MLCT excited states of Re(I), d-d excited states of Rh(III) and then to MLCT states of a series of Ru-bpy complexes[185]. In the case of ruthenium complexes there was an important development; three rate constants for return to the ground state could be delineated. In addition to $k_r$ and $k_{nr}$, which are essentially temperature independent, a new temperature-dependent term, $k'^{\circ}$, was found to be important. Thus, the observed rate constant in these cases was described as

$$k_{obs} = k_r + k_{nr} + k'^{\circ}\exp(-E_a/k_bT)$$

The temperature dependent term describes the transfer of CT energy to d-d excited states and constitutes an additional nonradiative deactivation pathway. (See Fig. 2.) Caspar and Meyer[185] calculated $E_a$ for $[Ru(bpy)_3]^{2+}$ to be 3560 $cm^{-1}$ supporting Van Houten and Watts original estimate of ca. 3600 $cm^{-1}$ [174].

Vibrational deactivation has two important components: trapping vibrations of the solvent (solvent interactions to be considered later in this paper) and metal-ligand vibra-

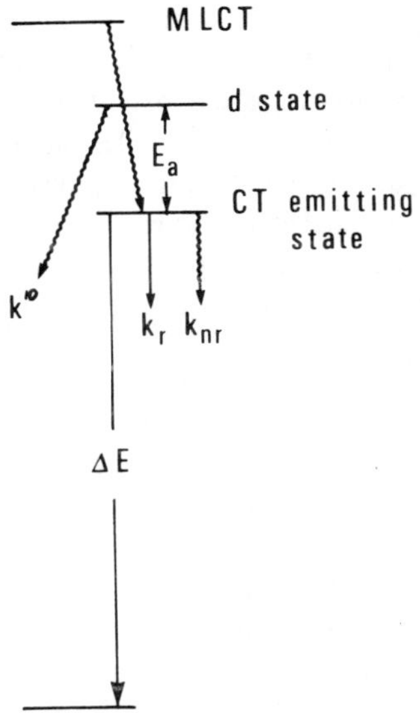


**Fig. 2.** Energy gap law diagram. ΔE is the energy gap (emission energy)

tional interactions. These latter deactivations were considered to come about from metal-nitrogen skeletal vibrations as promoting deactivation, and ligand skeletal vibrations (ring modes) as accepting the promoting energy. Values of $k_{nr}$ will increase with increasing frequency of the metal-nitrogen modes and decrease with increasing frequency of the ring modes.

The temperature dependent term, k'°, has important implications for ruthenium emission. An electron thermally excited from the MLCT into a d-d state is presumed to rapidly return to the ground state nonradiatively. If the MLCT state is of very low energy there should be a large $E_a$ to populate d-d and little thermal loss of emission. However, $k_{nr}$ would then be large from energy gap law considerations; the compound would be a poor emitter at all temperatures. If instead the MLCT state is of very high energy, the CT and d-d potential curves could cross near the minimum of the CT well and a pronounced temperature dependence would result. Caspar and Meyer argue[185)] that this is most likely the situation with $[Ru(bpy)_2(YR_3)_2]^{2+}$ (Y = P or As). These complexes emit only below 150 K, and $[Ru(bpy)_2(CO)Cl]^+$ is nonluminescent even at 77 K. The extreme photoreactivity of this complex[80)] is consistent with the presence of a d-d state below the lowest CT level. Thus, there is a rather crucial balance between the energies of d-d and MLCT states for efficient luminescence.

Kober and Meyer[186)] have compared osmium(II) bpy and phen complexes through calculations. These two ligands were found to be nearly equivalent as chromophores; their lowest π* orbitals are remarkably similar. They concluded that the observed differences in excited state properties may have more subtle origins than electronic effects.

Alford et al. concluded that the energy gap law does not hold in a series of phen complexes[25)]. The description of their lifetime equipment suggests that the instrument time should be acceptably short, but the lifetime given for $[Ru(bpy)_3]^{2+}$ is rather long (ca. 1.1 μs in ethanol-methanol at 20 °C). If some experimental error is giving long lifetimes for all of their compounds, then this conclusion is not valid or is it expected from Kober and Meyer's conclusions[186)].

We have seen from the above work that the nonradiative rate constants dominate the luminescence behavior of ruthenium(II) complexes. If one can increase the value of the radiative rate constant, $k_r$, without substantial increases in $k_{nr}$, then emission efficiency can be improved. The radiative rate constant is, in theory at least, related to the molar absorption coefficient, epsilon[187]. Demas and Crosby[188], made a number of assumptions and calculated radiative lifetimes based on observed epsilon values, which were in good agreement with the experimental $k_r$ values. Watts and Crosby[189] went on to comment on the possible implications of the epsilon value.

Cook et al.[7] prepared a number of ruthenium(II) complexes of substituted bpy's, and showed a reasonable correlation between $k_r$ and epsilon[190]. However, choosing a substituent to alter epsilon may not have the desired effect of increased efficiency. While $RuL_3^{2+}$ (L = 4-dimethylamino-bpy) has a larger value of epsilon than the parent complex, the emission energy is 1000 $cm^{-1}$ lower and the quantum yield is only ca. 30% that of $[Ru(bpy)_3]^{2+}$. They found the energy gap law to be followed in these compounds.

Some of Cook et al.'s compounds were anomalously efficient emitters at room temperature; $RuL_3^{2+}$ (L = 4,4′-diphenyl-bpy) has a quantum yield of 0.30 in alcoholic solution. Here they proposed a new concept in the model. Where substituents extend conjugation they propose the excited electron to be beyond a critical distance from the metal center. Being "far removed", nonradiative recombinations occur less frequently. In this fashion the nonradiative rate can be considerably decreased. This explanation possibly also covers $RuL_2^{2+}$ (L = 4,4′,4″-$ph_3$-trpy, ligand **46**) which has been observed to emit at room temperature[191] by Stone and Crosby. In contrast, $[Ru(trpy)_2]^{2+}$ is essentially nonemitting under these conditions[19]. In a following paper, Cook et al.[25] examined a large number of phen complexes. The concept of extended conjugation was borne out with 4,7-diaryl-substituted ligands giving complexes with the highest quantum yields, 4-aryl-substituted giving somewhat lower values, and all other substituents producing compounds with lower quantum yields.

Placing a phenyl in the 5-position results in a quantum yield only slightly greater than that of the unsubstituted parent complex (0.033 vs. 0.019). This is in keeping with the extended conjugation concept in that phen's 5 and 6 positions, while conjugated with each other, show little conjugation with the rest of the aromatic system. The apparent anomalies in this set of compounds are the chelates containing methyl groups in the 5 and 6 positions. A quantum yield here of 0.143 shows a much greater efficiency increase than in the bpy series, where the chelate with 4,4′-dimethyl-bpy is reported to have the same efficiency as the parent complex.

One final point of importance regards quenching by oxygen[7, 190]. Several years ago Demas and Adamson[173] concluded that nitrogen purging is adequate to reduce oxygen quenching to less than 1%. Much of the emission work reported to date has been done using this technique. However, Cook et. al.[7] claim the freeze-pump-thaw technique to be essential in studying these compounds. Using this method they obtain an emission quantum yield for $[Ru(bpy)_3]^{2+}$ in water of 0.062 as opposed to the accepted value of 0.042[174]. In their later paper[25] these workers commented that the earlier indications of higher quantum yields by this technique were not substantiated. Unfortunately, this places a large body of data in question[7, 190].

### *1. Solvent Effects*

There have been a number of reports on variation of observed lifetime and/or quantum yield with solvent. Here the necessity for careful reporting of methods becomes evident. Considerable variation between research groups exists for quantum yields of $[Ru(bpy)_3]^{2+}$ in the same solvent.

In Table 1 we have collected data on measured lifetimes and quantum yields in various solvents for $[Ru(bpy)_3]^{2+}$. Also included is a calculated quantum yield from the observed lifetime, assuming a constant value for the radiative lifetime (14 μs[174, 192]). Demas and Crosby[188] have argued that the intersystem crossing efficiency to populate the emitting MLCT triplet is unity and should be independent of solvent; Bolletta et al.[193]

**Table 1.** $[Rubpy_3]^{2+}$ quantum yields and lifetimes in solvents

| Solvent | T = 25 °C | | $\Phi_{calc}$ [a] | Ref. |
|---|---|---|---|---|
| | τ (μs) | Φ | | |
| $D_2O$ | 1.02 | 0.070 | 0.073 | 174 |
| $H_2O$ | 0.58 | 0.042 | | 174 |
| dichloromethane | 0.488 | 0.029 | 0.035 | 196 |
| n-butyronitrile | 0.918 | 0.060 | 0.066 | 196 |
| pyridine | 0.920 | 0.042 | 0.066 | 196 |
| acetonitrile | 0.855 | 0.062 | 0.061 | 196 |
| | | 0.075 | | 274 |
| propylene carbonate | 0.938 | 0.071 | 0.067 | 196 |
| DMF | 0.912 | 0.063 | 0.065 | 196 |
| $H_2O$ | 0.685 | 0.055 | 0.049 | 259 |
| ethanol | 0.900 | 0.078 | 0.064 | 259 |
| EPA | 0.945 | 0.130 | 0.068 | 259 |
| EPA | 0.926 | 0.270 (12 °C) | 0.066 | 260 |
| EtOH–MeOH (4 : 1) | 1.15 | 0.089 | 0.082 | 7 |
| $H_2O$ | | 0.062 | | 7 |
| glycerol | | 0.127 | | 7 |
| acetonitrile | | 0.122 | | 7 |
| DMSO | | 0.131 | | 7 |
| DMF | | 0.129 | | 7 |
| $H_2O$ | 0.588 | | 0.042 | 176 |
| $D_2O$ | 0.946 | | 0.068 | 176 |
| 9 MLiCl/Water | 0.666 | | 0.048 | 176 |
| 50% $H_2SO_4$ | 0.639 | | 0.046 | 176 |
| EtOH | 0.696 | | 0.050 | 176 |
| acetonitrile | 0.818 | | 0.058 | 176 |
| DMF | 0.851 | | 0.061 | 176 |
| DMSO | 1.02 | | 0.073 | 176 |
| cellulose acetate[b] | 1.47 | | 0.105 | 176 |
| polyvinyl alcohol[b] | 1.65 | | 0.118 | 261 |
| polyvinyl acetate[b] | | 0.237 | | 7 |

[a] $\Phi_{calc}$ based on $\tau_r$ of 14 μs

Thus: $\Phi_{calc} = \frac{\tau_{obs}}{14}$

[b] Polymer film

have found the same efficiency through quenching experiments. Van Houten and Watts found the radiant lifetime to be insensitive to solvent ($H_2O$ vs. $D_2O$)[194] and both temperature and solvent[174]. Further, Caspar et al.[195] have shown the radiant lifetime in $[Os(phen)_3]^{2+}$ to be insensitive to solvent and essentially unaffected by solvent in $[Ru(bpy)_3]^{2+}$[196]. However, some caution must be used in converting from observed lifetime to quantum yield. Bolletta et al.[197] have noted in cases where ion pairing occurs that lifetime and intensity quenching may not be parallel. The effect of counter ions on osmium(II) emission spectra has been considered by Meyer's group[198]. Yersin et al.[199] found the anion in a series of $[Ru(bpy)_3]X_2$ salts to have drastic effects on emission properties of the solids; this probably has less influence in solution work.

Caspar and Meyer[196] have made a detailed analysis of solvent effects on Ru-bpy emission. Several important points emerge from their analysis. First, it is only $k_{nr}$ which is influenced by solvent. Second, they argue that solvent effects can be simply reduced to energy gap effects; emission energy varies with solvent as does absorption energy. Meyer's group has taken this one step further[172] in demonstrating the solvent dependence of the MLCT absorption bands and relating it to a theoretical model. Earlier Allsopp et al. had shown the presence of two thermally activated decay routes[176], whose exact location is dependent on the medium.

An important aspect of the energy gap law is vibrational trapping of excited state energy through solvent acceptor vibrations. Cook et al.[7] point out that this is a reasonable cause for the higher quantum yields observed in $D_2O$ than in $H_2O$. Kurimura et al.[200] demonstrated that Ru-bpy in aqueous solutions of polystyrenesulfonate appears to be "encapsulated" by the soluble polymer and thus shielded from the solvent. This resulted in nearly a two-fold increase in emission intensity. Earlier Meyer's group[150] observed that polyvinylalcohol in water-ethylene glycol allowed the observation of emissions at room temperature in a system claimed to produce multiple emissions, and commented on the highly viscous nature of the solvent system. In light of more recent data, solvent viscosity does not appear to be the only deciding factor.

Earlier Van Houten and Watts[194] had concluded there to be some charge transfer to solvent (CTTS) character in the $[Ru(bpy)_3]^{2+}$ luminescent state. Nakamaru[201] recognized the importance of this, and found solvent dependence of $[Ru(bpy)_3]^{2+}$ emission correlated well with the Gutmann acceptor number, an empirical parameter quantifying a solvent's electrophilic properties[202]. In the range of solvents studied by Nakamaru, CTTS quenching accounted for 15% to 45% of the total nonradiative rate constant. A similar behavior with varying solvent was found in a mixed chelate[10]. Kawanishi et al.[203] felt Nakamaru's results may have been fortuitous and examined five different tris-chelates for solvent dependence. Correlation with the acceptor number was good in some cases and lacking in others. They commented that effects of ligand structure on emission properties are very complicated.

Belser et al.[204] studied the solvent dependence of the properties of several cyano-complexes, $[Ru(AB)_2(CN)_2]$. They noted a good correlation between absorption band energy and solvent acceptor number, the CT band energy increasing with this parameter. Decreased electron density at the metal center due to cyanide-solvent interactions was the explanation for their observations. Quantum yields were only reported in three solvents for this group of compounds[204] and no correlation with the acceptor number was apparent.

Since solvent vibrational modes appear to be one crucial factor in determining non-radiative decay, poorly solvating polymer films should be good media for efficient emission. Allsopp et al.[176] commented on this point, and the lifetime they observed in cellulose acetate film gives a room-temperature quantum yield of 0.10. The two-fold higher value observed by Cook et al. in polyvinylacetate (Table 1) may not be significant since this group's quantum yields tend to be higher.

Cook et al.[7] did compare results in several different polymer matrices for the chelate formed from perdeuterobipyridyl. Quantum yields in polyvinylchloride and polymethylmethacrylate varied significantly (0.356 vs. 0.230). They suggested the difference could arise from the lower density of high frequency vibrational modes in the former matrix, consistent with the conclusions developed by others.

Other rigid materials have been examined as matrices for the study of Ru-bpy emission. Gafney's group has investigated porous Vycor glass in this context and found the absorption and emission spectra of the absorbed complex are equivalent to aqueous solutions with comparable lifetime[205]. Here, anionic silanol groups are the binding sites and "solvent" effects are probably much like those of water. (An unusual observation was biphotonic excitation and emission with the complex in this matrix[206].) Ru-bpy has also been incorporated into zeolites[207] yielding properties similar to those of aqueous solutions, most likely for similar reasons.

### 2. *Other Deactivating Modes*

In addition to the factors considered above for nonradiative dissipation of excited state energy one might also assume heavy atom effects to be important. Demas and Addington[208] concluded both heavy atom effects and paramagnetic quenching to be unimportant pathways. However, in a set of $[Ru(bpy)_2(AA)]$ complexes Haga et al.[209] found no emission when a bidentate sulfur donor was present, while an analogous oxygen donor yielded weak emitters. But Root et al.[210] found $[Rubpy(trpy)L]^{2+}$ (L a sulfur donor) to be weak emitters at room temperature in aqueous solution. They noted that this ligand, when not coordinated, is an efficient quencher.

## B. *Summary of Emission Results*

### 1. *Introduction*

A few cases of emission have been reported for compounds of other types of ligands, which also deserve mention. The emission of $K_4[Ru(CN)_6]$ at 77 K[211,212] has been characterized as a d-d transition. Vogler and Kunkely have made a rare observation of CT emission from the ruthenium(III) compound, $K_3[Ru(CN)_6]$[213].

Ruthenocene has been observed to emit by Crosby's[214] and Wrighton's[215] groups. Both assign this low-temperature emission to a d-d excited state. Klassen and Crosby reported emission from a 1,2-(diphenylphosphino)ethane complex, $[Ru(LL)_2Cl_2]$[216]; however, DeArmond and co-workers[217] could not locate this emission. Ruthenium porphyrins appear to display ligand centered luminescence ($\pi, \pi^*$)[218,219]. Krug and Demas investigated the ruthenium(II) tris-chelate of 2,2'-bithiazoline[31] and observed a very weak, short-lived emission at 77 K. They concluded that an aromatic ligand is not essential for MLCT luminescence.

## 2. *Emission from Complexes of Bpy and Derivatives*

To assess the effect of ligand structure, several measurements are necessary: the emission energy, the lifetime, and the emission quantum yield. Furthermore, it is essential that details on precisely how the data have been obtained be given. Demas and Crosby[220] have pointed out the necessity for this reporting. In the absence of specific details, discrepancies between reported values are difficult to assess. From the data in the following sections it will be seen that there is a wide variation in the quantities reported.

Our procedure for tabulating emission data is to group complexes according to ligand type. This raises a question as to where mixed complexes are located. In the tables for bpy and derivatives are included mixed complexes with bpy if there is no emitting "parent" for the coligand; i.e., if the coligand does not appear in a separate section. The one exception to this procedure is mixed bpy/phen complexes, which are arbitrarily included with bpy. Homochelates of bpy and derivatives are tabulated in Table 2 while the mixed complexes appear in Tables 3, 4 and 5.

A number of the complexes in Table 2 have been investigated by several groups, and some discrepancies become apparent. Since the solvent affects emission kinetics, comparisons can only be made under identical conditions unless extensive data in different solvents are available. Consider the data for $[Ru(4,4'\text{-}ph_2\text{-}bpy)_3]^{2+}$: From the low-temperature data[142] a radiative lifetime of 8.6 μs is calculated. The room-temperature lifetime in water[19] gives an estimate of the quantum yield in this solvent of 0.078, compared with the reported value of 0.056. In room-temperature alcohol, the observed lifetime[7] gives a quantum yield estimate of 0.227 (0.306 reported). These discrepancies (ca. 30%) are much greater than found in $[Ru(bpy)_3]^{2+}$ (less than 10%).

A similar situation is observed with $[Ru(bpy)_2(4,4'\text{-}ph_2\text{-}bpy)]^{2+}$. The reported quantum yield at room temperature[7] does not agree well with that calculated from the low-temperature data[221] (20% discrepancy). It should be noted that Cook et al. originally felt the freeze-pump-thaw technique to be essential in quantum yield determinations, but later raised some doubt about the necessity for this practice[25]. There could have been some error in their methods leading to consistently high values.

Another type of discrepancy appears in $[Ru(4\text{-}NO_2\text{-}bpy)_3]^{2+}$. Gafney and co-workers[222] could detect no emission from this complex even at 77 K. A weak emission could be observed only when the complex was isolated in porous Vycor. However, the complex undergoes a slow solvolysis in aqueous ethanol, and the resulting species was found to be luminescent. This raises questions on the origin of the reported emission in this complex[223] and also on the chelate of $4,4'\text{-}NO_2\text{-}bpy$[7], which should undergo a similar thermal solvolysis reaction. In the mixed complexes of $5\text{-}NO_2\text{-}phen$[224], only a very weak emission could be detected. Resonance raman experiments[222] indicated that the excitation energy localizes in the nitro group of $4\text{-}NO_2\text{-}bpy$, which could easily be deactivated by energy transfer to the surroundings.

A number of studies have been reported on complexes of bpy derivatives containing hydrocarbon chains, esters or amides in the 4,4'-positions (in some cases, only mixed complexes with bpy were studied). These somewhat hydrophobic compounds were generally examined in terms of energy transfer. While it is difficult to compare results

**Table 2.** Complexes of substituted bipyridyls, $[Ru(AB)_3]^{2+}$

| Bpy substituent | Abs | | Em (kK) | | τ (μs) | | Solv. | Φ | | Ref. |
|---|---|---|---|---|---|---|---|---|---|---|
| | kK | E | 77 | RT | 77 | RT | | 77 | RT | |
| none | | | | | 5.21 | | alc. | | | 141 |
| | | | | | | | alc. | 0.376 | | 188 |
| | | | | | | 0.58 | $H_2O$ | | 0.042 | 174, 192 |
| | 22.0 | 14600 | | | | | | | | 42 |
| | | | 17.23 | | | | | | | 221 |
| | | | 15.88 | | | | | | | |
| | | | 14.69 | | | | | | | |
| | | | | 16.3[c] | | 0.60 | $H_2O$ | | | 19 |
| | | | | 15.9[c] | | | | | | |
| | | | | | | | $H_2O$(air) | | 0.029 | 192 |
| | | | | | | | $H_2O$ | | 0.062 | 7 |
| | | | | | | 1.15 | alc. | | | 7 |
| | 21.9 | 13000 | 16.7 | | 4.7 | | MF | | | 248 |
| | | | | | | 0.85 | eth | | 0.057 | 262 |
| | | | | 16.1[c] | | 0.850 | pC | | 0.040 | 27 |
| 4-Br | 21.9 | 14000 | | 15.6 | | 0.82 | alc | | 0.052 | 7 |
| 4-Cl | 22.0 | 13500 | | 15.6 | | 0.67 | alc | | 0.030 | 7 |
| 4-OMe | 21.7 | 14100 | | 15.2 | | 0.67 | alc | | 0.039 | 7 |
| 4-$OCH_2$-ph | 21.7 | 14800 | | 15.4 | | 0.60 | alc | | 0.053 | 7 |
| 4-$NH_2$ | 21.16 | 12100 | | 14.8 | | 0.35 | alc | | 0.021 | 7 |
| 4-N(Me) | 21.0 | 15000 | | 14.7 | | 0.35 | alc | | 0.023 | 7 |
| 4-$NO_2$ | 20.6 | | | 14.7 | | | glass | | | 222 |
| | 20.8 | 20400 | | 14.3 | | 0.12 | nitrile | | | 223 |
| | | | 15.6 | | 3.8 | | alc | | | 223 |
| 4,4′-($NO_2$) | 21.1 | – | | 14.3 | | 0.25 | alc | | 0.002 | 7 |
| 4,4′-($CO_2$Et) | 21.6 | 23300 | | 15.3 | | 1.65 | alc | | 0.200 | 7 |
| 4,4′-Br | 21.5 | 17300 | | 15.2 | | 0.52 | alc | | 0.060 | 7 |
| 4,4′-Cl | 21.5 | 15500 | | 14.9 | | 0.40 | alc | | 0.036 | 7 |
| | | | 16.7 | | 2.9 | | alc 77 | | | 223 |
| | 21.6 | 17000 | | 15.8 | | 0.48 | nitrile | | | |
| 4,4′-ph | 21.1 | 28000 | | 15.7 | | 1.95 | alc | | 0.306 | 7 |
| | | 32700 | | 15.7[c] | | 0.67 | $H_2O$ | | | 19 |
| | | | | | 4.9 | | alc | 0.57 | | 142 |
| | | | 16.4 | | | | alc | | | 189 |
| | | | | | | | $H_2O$ | | 0.056 | 192 |
| 4,4′-(C=C-ph) | 20.5 | 33000 | | 14.7 | | 0.72 | alc | | 0.030 | 7 |
| 4,4′-($CH_2$-ph) | 21.7 | 16100 | | 15.6 | | 1.25 | alc | | 0.098 | 7 |
| 4,4′-(Me) | 22.0 | 17000 | | 15.6 | | 0.95 | alc | | 0.086 | 7 |
| | 21.7 | 14300 | | 15.8[c] | | 0.33 | $H_2O$ | | | 19 |
| | | | | | 4.6 | | alc | 0.283 | 0.026 | 10 |
| | | | | | | | | | $H_2O$ | |
| | | | | | | | $H_2O$ | | 0.022 | 192 |
| 6-Me | 22.3 | 11100 | | | | | MeOH | | | 42 |
| | | | 17.1 | | 4.1 | | alc | 0.0968 | | 42 |
| 6,6′-(Me) | 22.4 | 7440 | | | | | MeOH | | | 42 |
| | | | 17.0 | | 2.5 | | alc | 0.0176 | | 42 |
| 5,5′-(Me) | 22.7 | 14700 | | 16.1 | | 0.35 | alc | | 0.037 | 7 |
| 4,4′-(O-ph) | 20.9 | 13400 | | 14.9 | | 0.35 | alc | | 0.038 | 7 |
| 4,4′-(OEt) | 21.0 | 13000 | | 14.8 | | 0.30 | alc | | 0.020 | 7 |
| 4,4′-($OCH_2$-ph) | 21.0 | 13500 | | 14.9 | | 0.31 | alc | | 0.032 | 7 |
| 4,4′-(NHAc) | 20.9 | 15000 | | 14.9 | | 0.41 | alc | | 0.027 | 7 |

**Table 2** (continued)

| Bpy substituent | Abs | | Em (kK) | | τ (μs) | | Solv. | Φ | | Ref. |
|---|---|---|---|---|---|---|---|---|---|---|
| | kK | E | 77 | RT | 77 | RT | | 77 | RT | |
| 4,4′-($NH_2$) | 19.8 | 10500 | | 14.2 | | 0.10 | alc | | 0.004 | 7 |
| 4,4′-($NEt_2$) | 19.3 | 14500 | | 14.3 | | 0.13 | alc | | 0.010 | 7 |
| 5,5′-($CO_2Et$) | 20.2 | 9900 | | 13.9 | | 0.23 | alc | | 0.004 | 7 |
| 5,5′-(NHAc) | 22.5 | 14900 | | 15.9 | | 2.40 | alc | | 0.126 | 7 |
| 3,3′-(Me) | | | 17.0 | | 6.4 | | alc | 0.113 | | 10 |
| | 21.9 | 12600 | | 15.8 | | | $H_2O$ | | | 10 |
| | 21.9 | 10830 | | 16.0 | | 0.21 | nitrile | | | 249 |
| | | | 16.8 | | | | alc | | | 249 |
| 4,4′-(t-Bu) | 21.9 | 16800 | | 16.0 | | 1.15 | nitrile | | | 249 |
| | | | 17.4 | | | | alc | | | 249 |
| 4-($Et_3P^+$) | 21.5 | 18300 | | 15.9 | | | $H_2O$ | | | 222 |
| 4,4′-ester 13a | | | | 15.9 | | 2.39 | nitrile | | | 263 |
| 4,4′-ester 13b | | | | 15.9 | | 2.21 | nitrile | | | 263 |
| 4,4′-ester 13c | | | | 15.7 | | 1.93 | nitrile | | | 263 |
| 4,4′-ester 13d | | | | 15.8 | | 2.19 | nitrile | | | 263 |
| 4,4′-ester 13e | | | | 15.8 | | 2.21 | nitrile | | | 263 |
| 4,4′-ester 13f | | | | 15.7 | | 2.14 | nitrile | | | 263 |
| 4,4′-$CO_2Me$ | | | | | | 0.89 | $H_2O$ | | | 44 |
| 4,4′-$CO_2^-$ | | | | | | 0.62 | $H_2O$ | | | 44 |
| 4,4′-$SO_3^-$ | 20.0 | | | | | 0.48 | $H_2O$ | | | 44 |
| 4,4′-$SO_3^-$ | | | | | | 0.87 | MeOH | | | 44 |
| 4,4′-$SO_3^-$ | | | | | | 0.99 | $D_2O$ | | | 44 |
| Ligand 15 | 20.0 | 19500 | | 13.4 | | | nitrile | | | 239 |
| | | | 13.6 | | 1.1 | | alc | | | 239 |
| Ligand 16 | 22.0 | 14700 | | 16.4 | | | nitrile | | | 239 |
| | | | 17.2 | | 5.5 | | alc | | | 239 |

coming from different groups, some interesting conclusions can be reached. Ford and Calvin[225] noted an amide on bpy gave rise to a very marked solvent dependence in both emission energy and absorbance band shape, which was attributed to solvent polarity affecting the energy for transfer of an electron to the substituted bpy. Ferguson et al.[226] concluded that electron withdrawing carbonyl groups on a substituted bpy (mixed complex) caused the excited electron to localize on that ligand. Johansen et al.[227] suggested from energy-transfer experiments that alkyl substituents on bpy take part in transfer of excitation energy.

The cyano-complex, $[Ru(bpy)_2(CN)_2]$, is quite sensitive to solvent in both absorption and emission[228–230] undoubtedly because of the basic, uncoordinated cyano-nitrogens. Demas and co-workers have shown that on protonation the absorbance shifts to higher energy and the protonated compound is nonemitting[230]. Further, since the protonated excited state is a stronger acid than the ground state complex[231], they argued that in $[Ru(phen)_2(CNH)_2]^{2+}$ the observed emission is ligand phosphorescence, with the MLCT state being raised above the lowest ligand localized $^3(\pi\text{-}\pi^*)$. Such an electronic situation may also exist for the bpy analogue.

**Table 3.** Mixed complexes $[Ru(bpy)_2L_n]^{x+}$

| Ligand | (n) | Abs | | Em (kK) | | τ (μs) | | Solv. | Φ | | Ref. |
|---|---|---|---|---|---|---|---|---|---|---|---|
| | | kK | E | 77 | RT | 77 | RT | | 77 | RT | |
| 1 | (1) | 22.2 | 10800 | | 16.6 | | | nitrile | | 4 × $10^{-3}$ | 21 |
| 2 | (1) | 21.8 | 14600 | | 15.3 | | | nitrile | | | 11 |
| cis-py | (2) | | | 17.2 | | 11.4 | | gly | | | 146 |
| | | 22.2 | 7900 | | | | | nitrile | | | 93 |
| cis-py | (2) | | | | | | 0.06 | $H_2O$ | | | 265 |
| cis-py | (2) | 21.9 | | | 16.5 | | 0.500 | nitrile | | | 266 |
| trans-py | (2) | | | 17.1 | | 10.6 | | gly | | | 146 |
| | | 21.1 | 8600 | | | | | nitrile | | | 93 |
| phen | (1) | | | 17.40 | | 6.6 | | alc | 0.44 | | 221 |
| | | | | | | | 0.684 | $H_2O$ | | | 224 |
| 5,6-Mephen | (1) | | | 17.32 | | 6.1 | | alc | 0.42 | | 221 |
| 4,7-Mephen | (1) | | | 17.15 | | 5.6 | | alc | 0.38 | | 221 |
| 2,9-Mephen | (1) | 22.1 | 13600 | | 17.0 | | | nitrile | | | 236 |
| 4,7-ph-phen | (1) | | | 16.98 | | 9.4 | | alc | 0.54 | | 221 |
| | | | | | | | 1.591 | $H_2O$ | | | 224 |
| ox | (1) | 19.1(l) | | 15.0 | | | | alc | | | 140 |
| | | | | | | 0.61 | | alc | 0.0124 | | 188 |
| en | (1) | 19.8(l) | | 15.5 | | | | alc | | | 140 |
| | | | | | | 0.96 | | alc | 0.0222 | | 188 |
| cis Cl | (2) | 19.2(l) | | 15.2 | | | | alc | | | 140 |
| | | 18.6 | 9890 | 13.3 | | 0.17 | | MF | | | 248 |
| 3 | (1) | 18.8 | 27300 | | 16.3 | | | nitrile | | | 264 |
| 5-$NO_2$phen | (1) | | | | | | 0.007 | $H_2O$ | | | 224 |
| H19 | (1) | 21.7 | 12600 | | 15.34 | | 0.455 | $H_2O$ | | 0.01 | 54 |
| | | | | 16.75 | | 3.90 | | eth. | 0.27 | | 54 |
| 19 | (1) | 20.2 | 10000 | | 13.70 | | 0.038 | aq | | 0.001 | 54 |
| 4 | (1) | 21.8 | 12900 | | | | | nitrile | | | 240 |
| | | | | | 15.9 | 4.24 | 0.262 | aq | | | 240 |
| | | | | 16.7 | 16.1 | | 0.418 | alc | | | 240 |
| 4-H | (1) | | | | 14.5 | | | aq | | | 240 |
| | | 20.3 | 9500 | | | | | nitrile | | | 240 |
| 5 | (1) | 21.7 | 10500 | | | | | nitrile | | | 240 |
| | | | | | 15.8 | | 0.224 | aq | | | 240 |
| | | | | 16.8 | 16.0 | 4.45 | 0.160 | alc | | | 240 |
| 5-H | (1) | | | | 14.5 | | | aq | | | 240 |
| | | 20.0 | 9100 | | | | | nitrile | | | 240 |
| 6 | (1) | 21.6 | 11700 | | | | | nitrile | | | 240 |
| | | | | | 15.6 | | 0.111 | aq | | | 240 |
| | | | | 16.2 | 15.6 | 3.00 | 0.120 | alc | | | 240 |
| 6-H | (1) | | | | 14.2 | | | aq | | | 240 |
| | | 19.7 | 9100 | | | | | nitrile | | | 240 |
| 7 | (1) | 21.1 | 9300 | | | | | nitrile | | | 240 |
| | | | | | 15.2 | | 0.092 | aq | | | 240 |
| | | | | 16.1 | 15.4 | 3.25 | 0.150 | alc | | | 240 |
| 7-H | (1) | 19.5 | 9500 | | | | | nitrile | | | 240 |
| | | | | | 14.2 | | | aq | | | 240 |
| 7 | (1) | 22.3 | 9500 | | | | | aq | | | 236 |
| | | | | | 16.8 | | | nitrile | | | 236 |
| H-8 | (2) | 21.3 | 8700 | | | | | nitrile | | | 110 |
| | | | | | 15.1 | | | ac | | | 110 |

**Table 3** (continued)

| Ligand | (n) | Abs kK | Abs E | Em (kK) 77 | Em (kK) RT | τ (μs) 77 | τ (μs) RT | Solv. | Φ 77 | Φ RT | Ref. |
|---|---|---|---|---|---|---|---|---|---|---|---|
| ph-8 | (2) | 21.4 | 8700 | | | | | nitrile | | | 110 |
| | | | | | 15.5 | | | ac | | | 110 |
| $C_3H_5$-8 | (2) | 21.2 | 9300 | | | | | nitrile | | | 110 |
| | | | | | 15.2 | | | ac | | | 110 |
| Me-8 | (2) | 21.1 | 8900 | | | | | nitrile | | | 110 |
| | | | | | 15.2 | | | ac | | | 110 |
| 9 | (2) | 21.3 | 5100 | | 15.9 | | | nitrile | | | 110 |
| 10a | (1) | 19.5 | 9300 | | | | | nitrile | | | 209 |
| | | 19.5 | | 14.2 | 13.6 | | | alc | | $10^{-6}$ | 209 |
| 10b | (1) | 19.6 | 9500 | | | | | nitrile | | | 209 |
| | | 19.4 | | 14.8 | 13.5 | | | alc | | $10^{-6}$ | 209 |
| 10c | (1) | 20.2 | 7400 | | | | | nitrile | | | 209 |
| | | 19.9 | | 14.3 | | | | alc | | | 209 |
| 10d | (1) | 20.7 | 14000 | | | | | nitrile | | | 209 |
| | | 20.2 | | 14.8 | 14.7 | | | alc | | $10^{-6}$ | 209 |
| 10e | (1) | 21.3 | 11000 | | | | | nitrile | | | 209 |
| | | 21.5 | | 14.4 | | | | alc | | | 209 |
| 11 | (1) | 23.3 | 12000 | | | | | | | | 245 |
| | | 21.3 | sh | | 14.8 | | 0.135 | eth | | 0.05 | 245 |
| 11 | (1/2) | 19.0 | 21000 | | 13.2 | | 0.054 | eth | | 0.003 | 245 |
| 12 | (1) | 19.0 | 8490 | | 13.0 | | | nitrile | | ~6 × $10^{-6}$ | 267 |
| 12 | (1/2) | 18.9 | 16300 | | 12.8 | | | nitrile | | ~6 × $10^{-6}$ | 267 |
| CN | (2) | 21.19(1) | 9810 | 17.12 | | 3.96 | | alc | 0.27 | | 188 |
| | | | | | | | 0.27 | $H_2O$ | | | 228 |
| | | | | | | | 0.40 | MeOH | | | 268 |
| CN | (2) | 23.4 | | | 16.1 | | 0.250 | $H_2O$ | | | 233 |
| | | 19.8 | | | 14.7 | | 0.205 | DMF | | | 229 |
| 14 | (1) | | | 17.4 | | 5.9 | | alc | | | 239 |
| | | 22.2 | 13700 | | 15.2 | | | nitrile | | | 239 |
| 17 | (1) | 17.8 | 8540 | | 13.2 | | | nitrile | | | 239 |
| | | | | 13.5 | | 2.6 | | alc | | | 239 |
| $(CN)_2(Ag)_2$ | | 20.2 | | | 14.3 | | | nitrile | | | 230 |
| | | 22.9 | | | 15.4 | | | nitrile | | | 230 |
| $(CN)_2(Pt1)_2$ | | 22.2 | | | 16.1 | | 0.640 | DMF | | | 229 |
| $(CN)_2(Pt2)$ | | 24.0 | | | 16.9 | | 0.140 | $H_2O$ | | | 233 |
| | | 21.7 | | | 15.9 | | 0.630 | DMF | | | 233 |
| $(CN)_2(Pt2)_2$ | | 24.5 | | | 17.2 | | 0.060 | $H_2O$ | | | 233 |
| | | 23.5 | | | 17.2 | | 0.090 | DMF | | | 233 |
| 29 | (1) | 22.6 | 10900 | | 16.7 | | | nitrile | | | 236 |
| 44 | (1) | 20.2 | 14000 | | | | | MeOH | | | 181 |
| | | | | 14.7 | | 0.878 | | alc | | | 181 |

Demas et al.[228] observed luminescence quenching of $[Ru(bpy)_2(CN)_2]$ by aquo ions of cobalt(II), nickel(II), and copper(II) as well as by a number of other complexes. However, Kinnaird and Whitten[232] observed blue shifts in both absorbance and emission and substantial emission enhancement in a silver(I) adduct, $[Ru(bpy)_2(CN)(CNAg)]^+$,

**Table 4.** Mixed complexes, $[Ru(bpy)L_n]^{x+}$

| Ligands | (n) | Abs | | Em (kK) | | τ (μs) | | Solv. | Φ | | Ref. |
|---|---|---|---|---|---|---|---|---|---|---|---|
| | | kK | E | 77 | RT | 77 | RT | | 77 | RT | |
| 1 | (2) | 23.0 | 12200 | | 16.6 | | | nitrile | | $10^{-3}$ | 21 |
| phen | (2) | | | 17.4 | | 15.8 | | gly | | | 146 |
| | | 22.6 | 18000 | | | | | nitrile | | | 93 |
| | | | | 17.5 | | | | alc | 0.57 | | 145 |
| phen | (2) | | | 17.56 | | 9.1 | | alc | 0.56 | | 221 |
| | | | | | | | 0.784 | $H_2O$ | | | 224 |
| py | (4) | | | 17.1 | | 10.9 | | gly | | | 146 |
| | | 21.9 | 5500 | | | | | nitrile | | | 93 |
| 4,4′-ph-bpy | (2) | 21.5 | 14700 | | 15.7 | | 2.12 | alc | | 0.098 | 7 |
| 4,7-ph-phen | (2) | | | | | | 2.646 | $H_2O$ | | | 224 |
| 5-$NO_2$-phen | (2) | | | | | | 0.007 | $H_2O$ | | | 224 |
| 13f | (2) | | | | 15.5 | | 2.10 | nitrile | | | 263 |
| 4,4′-$CO_2$-bpy | (2) | | | | | | 0.54 | $H_2O$ | | | 44 |
| 4,4′-$SO_3$-bpy | (2) | 21.1 | | | | | 0.46 | $H_2O$ | | | 44 |
| 14 | (2) | 22.4 | 11300 | | 16.4 | | | nitrile | | | 239 |
| | | | | 17.4 | | 5.2 | | alc | | | |

which appeared to retain its integrity on excitation. Similarly, Bartocci et al.[229] observed spectral blue shifts and emission enhancement when five different platinum(II) complexes were substituted onto the uncoordinated nitrogen of $[Ru(bpy)_2(CN)_2]$. In a following paper[233] Bignozzi and Scandola found the 1:1 adduct with diethylenetriamine-platinum(II) to have enhanced emission, while attachment at both cyanides gave the same emission intensity as observed in $[Ru(bpy)_2(CN)_2]$. They also found the polymeric cations, $[Ru(bpy)_2CN]_n^+$, to emit in solution[234].

In a study of Fe–CN complexes, Shriver and Posner[235] concluded that an electron-withdrawing substituent on the nitrogen end of cyanide will increase the pi-acceptor character of this ligand. It is possible that attachment of groups to uncoordinated potential ligators on a complex may allow both tuning of emission energies and increase in efficiency. It will be interesting to determine which groups enhance or quench emission.

A group of ligands rarely encountered in ruthenium chemistry are the Schiff bases formed by condensation of an aromatic heterocyclic aldehyde and an amine. This scarcity is most likely due to ligand instability leading to complex mixtures on reaction with the common ruthenium source materials. However, Belser and von Zelewsky[11] have reported $[Ru(bpy)_2(\mathbf{2})]^{2+}$, which emits in solution. On further development of synthetic methods, it may be possible to obtain a variety of new luminescent compounds based on the Schiff bases. Dose and Wilson[236] reported blue emission from a tris-chelate of a Schiff base, but questioned the results. In our hands, the reaction mixture displays a blue luminescence, but on thin-layer chromatography of the crude mixture all emission originates from a colorless component, which is apparently an organic byproduct.

One final comment on these data emphasizes the sensitive nature of the luminescent measurement. Several groups have reported room-temperature solution emission from *cis*-$[Ru(bpy)_2(py)_2]^{2+}$, but Cherry's group reports this complex to contain a small amount of an emitting impurity (probably $[Ru(bpy)_3]^{2+}$)[237]. It is only this impurity which is

**Table 5.** Mixed complexes, $[Ru(bpy)_2(bpy\text{-}R)]^{2+}$

| Substituent, R | Abs | | Em (kK) | | τ (μs) | | Solv. | Φ | | Ref. |
|---|---|---|---|---|---|---|---|---|---|---|
| | kK | E | 77 | RT | 77 | RT | | 77 | RT | |
| 4,4′-Me | | | 17.06 | | 5.2 | | alc | 0.35 | | 221 |
| 4,4′-ph | | | 16.85 | | 5.6 | | alc | 0.46 | | 221 |
| 4,4′-ph | 21.9 | 19700 | | 15.9 | | 1.92 | alc | | 0.197 | 7 |
| 4,4′-($CO_2C_{18}H_{37}$) | 20.7 | | | 15.0 | | 0.900 | $CHCl_3$ | | 0.180 | 259 |
| | | | | | | 0.98 | eth | | 0.082 | 262 |
| 4,4′-($CO_2Me$) | 21.0 | | | 15.2 | | 0.615 | $H_2O$ | | 0.045 | 259 |
| 4,4′-($CO_2H$) | | | | 14.2 | | 0.274 | aq | | 0.013 | 269 |
| 4($CO_2H$)/4′($CO_2^-$) | | | | 14.7 | | 0.288 | aq | | 0.019 | 269 |
| 4,4′($CO_2^-$) | | | | 15.2 | | 0.555 | aq | | 0.037 | 269 |
| 4($CO_2H$)/4′-Me | | | | 13.9 | | 0.120 | aq | | 0.006 | 269 |
| 4($CO_2^-$)/4′-Me | | | | 15.4 | | 0.590 | aq | | 0.040 | 269 |
| 6-(p-tolyl) | 22.2 | 13500 | | 16.2 | | | ac | | wk | 270 |
| 6-(p-styryl) | 22.2 | 13800 | | 16.2 | | | ac | | wk | 270 |
| 4,4′-(CONH $C_{16}H_{33}$) | 21.7 | 14000 | | 16.3 | | | $CHCl_3$ | | | 225 |
| | 20.6 | 12000 | | 15.2 | | | aq | | | 225 |
| ester 13f | | | | 15.2 | | 2.00 | nitrile | | | 263 |
| 4,4′-alkyl | | | | | | 0.85 | eth | | 0.06 | 262 |
| 4,4′-($COC_{18}H_{37}$) | 21.4 | | | 14.3 | | 0.88 | eth | | 0.06 | 262 |
| 4,4′-$CO_2^-$ | | | | | | 0.53 | pH 3.5 | | | 44 |
| 4,4′-$CO_2^-$ | | | | | | 0.50 | pH 7.4 | | | 44 |
| 4-$NO_2$ | 20.3 | 11100 | 15.3 | 16.0 | | 0.78 | nitrile | | | 223, 239 |
| | | | 15.3 | | 3.1 | | alc | | | |
| 4,4′-Cl | 22.3 | 12600 | | 15.5 | | 0.42 | nitrile | | | 223 |
| | | | 16.4 | | 2.6 | | alc | | | |
| 15 | 20.6 | 14600 | | 13.8 | | | nitrile | | | 239 |
| | | | 15.2 | | 2.3 | | alc | | | |
| 16 | 22.1 | 13900 | | 16.3 | | | nitrile | | | 239 |
| | | | 17.1 | | 4.9 | | alc | | | 239 |
| 4,4′-$NO_2$ | 19.3 | 10300 | | 13.3 | | | nitrile | | | 239 |
| | | | 14.3 | | 2.5 | | alc | | | 239 |
| 18 | 19.9 | 11000 | | 13.9 | | | nitrile | | | 239 |
| | | | 16.1 | | 1.7 | | alc | | | 239 |

responsible for emission substantially above 77 K. Similar situations have probably been the source of many room-temperature observations.

## 3. *Emission from Complexes of Phen and Derivatives*

Comparison of the data in Tables 6 and 2 shows that the tris-chelates with phen's or bpy's are quite similar. For example, quantum yields for the bpy and phen complexes in water are similar (0.042 vs. 0.058)[192] – a similarity which persists whenever direct comparisons can be made. In most cases the phen chelate has an ε value ca. 25% higher than the corresponding bpy compound and the quantum yields are higher. (Unfortunately, few comparisons can actually be made because of the discrepancies noted earlier.)

Again in the phen[25] series the concept of extension of the conjugated system[190] as a means of increasing luminescence efficiency is borne out. Alford et al.[25] have presented

**Table 6.** Complexes of phen or derivatives $[Ru(phen)_3]^{x+}$

| Ligand | Abs | | Em (kK) | | τ (μs) | | Solv. | Φ | | Ref. |
|---|---|---|---|---|---|---|---|---|---|---|
| | kK | E | 77 | RT | 77 | RT | | 77 | RT | |
| phen | | | 17.7 | | 9.93 | | EMPA | | | 271 |
| | 24.6 | 19000 | | | | | aq | | | 271 |
| | | | | | 9.79 | | alc | 0.584 | | 272 |
| | 22.4 | 19000 | | | | | MeOH | | | 42 |
| | | | 17.6 | | 9.8 | | alc | 0.600 | | 42 |
| | 22.4 | | | 16.6 | | | $H_2O$ | | 0.058 | 192 |
| | 22.4 | 19000 | | 16.5[c] | | 0.92 | $H_2O$ | | | 19 |
| | | | | | | 0.89 | $H_2O$ | | | 176 |
| | | | | | | 1.09 | $D_2O$ | | | 176 |
| | | | | | | 0.22 | eth | | | 176 |
| | | | | | | 1.05 | DMSO | | | 176 |
| | 22.5 | 20000 | | 16.8 | | 0.45 | alc | | 0.019 | 25 |
| | | | | | | 0.908 | $H_2O$ | | | 273 |
| | | | | | | 1.09 | $D_2O$ | | | 273 |
| | | | | | | 0.313 | MeOH | | | 268 |
| 4,7-ph- | | | 17 | | 9.58 | | alc | 0.682 | | 272 |
| | 21.6 | | | 16.4 | | | eth | | 0.28 | 192 |
| | 21.7 | 29500 | | 16.3[c] | | 4.68 | $H_2O$ | | | 19 |
| | 21.6 | 28600 | | 16.2 | | 6.40 | alc | | 0.366 | 25 |
| | | | | | | 3.58 | $H_2O$ | | | 273 |
| | | | | | | 5.34 | MeOH | | | 268 |
| | | | 16.8 | | | | alc | | | 189 |
| 2,9-$CH_3$ | 20.0 | 2580 | | | | | MeOH | | | 42 |
| | | | 17.0 | | 2.3 | | | 0.0479 | | 42 |
| 2-$CH_3$ | 22.4 | 19000 | | | | | MeOH | | | 42 |
| | | | 17.6 | | | | alc | 0.172 | | 42 |
| 5-Cl | 22.4 | 17900 | | 16.6 | | 0.67 | alc | | 0.035 | 25 |
| | 22.3 | | | 16.6 | | | $H_2O$ | | 0.056 | 192 |
| | 22.4 | 18400 | | 16.5[c] | | 0.94 | $H_2O$ | | | 19 |
| 5-Br | 22.3 | | | 16.6 | | | $H_2O$ | | 0.069 | 192 |
| | 22.3 | 18800 | | 16.5[c] | | 1.04 | $H_2O$ | | | 19 |
| | | | | | | 1.21 | $H_2O$ | | | 273 |
| 4,7-$CH_3$ | 22.5 | 25300 | | 16.3[c] | | 1.74 | $H_2O$ | | | 19 |
| | 22.3 | 22700 | | 16.7 | | 0.85 | alc | | 0.059 | 25 |
| 5,6-$CH_3$ | 22.1 | 20400 | | 16.4[c] | | 1.81 | $H_2O$ | | | 19 |
| | 22.2 | 19800 | | 16.7 | | 2.50 | alc | | 0.143 | 25 |
| 5-$CH_3$ | 22.2 | 19400 | | 16.5[c] | | 1.33 | $H_2O$ | | | 19 |
| | 22.5 | 21000 | | 16.8 | | 0.85 | alc | | 0.061 | 25 |
| 3,4,7,8-$CH_3$ | 22.8 | 24500 | | 16.5[c] | | 1.39 | $H_2O$ | | | 19 |
| | 22.8 | 23300 | | 16.8 | | 0.48 | alc | | 0.032 | 25 |
| 3,5,6,8-$CH_3$ | 22.7 | 19800 | | 16.5 | | 2.22 | $H_2O$ | | | 19 |
| 5-ph- | 22.3 | 24600 | | 16.5 | | 1.29 | $H_2O$ | | | 19 |
| | 22.3 | 21700 | | 16.8 | | 1.15 | alc | | 0.033 | 25 |
| 5-$NO_2$- | 22.3 | 20000 | | ~16.5 | | $<$.005 | $H_2O$ | | | 19 |
| 4-$CH_3$ | 22.5 | 22000 | | 16.6 | | 0.80 | alc | | 0.054 | 25 |
| 4-Cl | 22.2 | 18100 | | 15.9 | | 1.78 | alc | | 0.037 | 25 |
| 4,7-Cl | 21.9 | 20600 | | 15.0 | | 1.55 | alc | | 0.028 | 25 |
| | 21.9 | 22000 | | 15.9[c] | | 2.5 | nitrile | | | 9 |
| 4-ph- | 22.2 | 22100 | | 16.5 | | 4.00 | alc | | 0.109 | 25 |
| 4-p-tolyl- | 22.1 | 23600 | | 16.5 | | 3.50 | alc | | 0.117 | 25 |
| 4-p-ph-ph- | 22.0 | 28700 | | 16.3 | | 4.90 | alc | | 0.184 | 25 |

**Table 6** (continued)

| Ligand | Abs | | Em (kK) | | τ (μs) | | Solv. | Φ | | Ref. |
|---|---|---|---|---|---|---|---|---|---|---|
| | kK | E | 77 | RT | 77 | RT | | 77 | RT | |
| 4-p-Br-ph | 22.1 | 25300 | | 16.3 | | 4.10 | alc | | 0.136 | 25 |
| 4-p-MeO-ph | 22.1 | 25300 | | 16.3 | | 3.25 | alc | | 0.130 | 25 |
| 4,7-p-tolyl | 21.6 | 27400 | | 16.2 | | 6.43 | alc | | 0.280 | 25 |
| 4,7-p-ph-ph | 21.5 | 36300 | | 16.1 | | 7.20 | alc | | 0.360 | 25 |
| 4,7-pBr-ph | 21.6 | 33000 | | 16.0 | | 5.95 | alc | | 0.403 | 25 |
| 4,7-pMeO-ph | 21.4 | 37900 | | 16.3 | | 6.10 | alc | | 0.387 | 25 |
| 4-p-ph-ph,7-ph | 21.5 | 33700 | | 16.2 | | 6.10 | alc | | 0.330 | 25 |
| 4-p-ph-ph, 7-pBr-ph | 21.5 | 31000 | | 16.0 | | 6.00 | alc | | 0.310 | 25 |
| 4-p-ph-ph, 7-pMeO-ph | 21.4 | 34800 | | 16.1 | | 6.05 | alc | | 0.305 | 25 |
| 4-pBr-ph, 7p-MeO-ph | 21.6 | 25500 | | 16.0 | | 5.95 | alc | | 0.237 | 25 |
| 5-Cl | | | | | | 0.962 | $H_2O$ | | | 273 |
| 5-$CH_3$ | | | | | | 1.33 | $H_2O$ | | | 273 |
| 4,7-$CH_3$ | | | | | | 1.65 | $H_2O$ | | | 273 |
| 5,6-$CH_3$ | | | | | | 1.91 | $H_2O$ | | | 273 |
| 3,4,7,8-$CH_3$ | | | | | | 1.75 | $H_2O$ | | | 273 |
| 5-ph | | | | | | 1.63 | $H_2O$ | | | 273 |
| 21 | 19.9 | 50000 | | <14[c] | | 1.4 | nitrile | | | 9 |

the largest collection of data yet reported on phen compounds. The nine phen complexes with the lowest $k_{nr}$ values in this study bear aromatic groups in the 4,7-positions, supporting the concept of extended conjugation.

An interesting point arises in a comparison of quantum yields for the 4,7-diphenyl-phen complex (0.366)[25] and the 4,4′-diphenyl-bpy compound (0.306)[7]. In phen the 4,7-aromatic substituents should not lie in the same plane as the remainder of the ligand and the extent of conjugation should be somewhat less than in the 4,4′-diphenyl-bpy ligand. Yet the increase of efficiency over the unsubstituted parent complex is far greater in the phen compounds.

One complex, the tris-chelate of 4,4′-distyryl-bpy, was anomalous with very poor emission efficiency, contrary to expectations. While no rationalization was offered, we

**Table 7.** Mixed complexes with phen, $[Ru(phen)_2L_n]^{x+}$

| Ligand | (n) | Abs | | Em (kK) | | τ (μs) | | Solv. | Φ | | Ref. |
|---|---|---|---|---|---|---|---|---|---|---|---|
| | | kK | E | 77 | RT | 77 | RT | | 77 | RT | |
| CN | (2) | | | | | | 0.654 | $H_2O$ | | | 273 |
| | | | | | | | 1.58 | MeOH | | | 268 |
| | | 22.2 | 10000 | | 15.7 | | | MeOH | | | 228 |
| | | | | 17.7 | | | | aq. MeOH | | | 228 |
| PDP(20) | (1) | 22.7sh | 19000 | | 16.7 | | 0.53 | nitrile | | | 9 |

**Table 8.** Mixed complexes of phen, $[Ru(phen)_2(phen')]^{x+}$

| phen' | Abs | | Em (kK) | | τ (μs) | | Solv. | Φ | | Ref. |
|---|---|---|---|---|---|---|---|---|---|---|
| | kK | E | 77 | RT | 77 | RT | | 77 | RT | |
| 4,7-ph | | | | | | 3.07 | $H_2O$ | | | 273 |
| | | | | | | 2.56 | MeOH | | | 268 |
| 5-Cl | | | | | | 0.947 | MeOH | | | 268 |
| 5-Br | | | | | | 0.989 | MeOH | | | 268 |
| H19 | 22.0 | 15000 | | 15.75 | | 1.250 | $H_2O$ | | 0.04 | 54 |
| 19 | 20.4 | 12380 | | 13.91 | | 0.175 | $H_2O$ | | 0.003 | 54 |
| 4,7-Cl | 22.8 | 18000 | | 15.5[c] | | 2.2 | nitrile | | | 9 |
| 20 | ca. 22.7 | 19000 | | 16.7 | | 0.53 | nitrile | | | 9 |
| 21 | 21.6 | 44000 | | 16.1[c] | | 1.4 | nitrile | | | 9 |
| H21 | 21.5 | 11000 | | 15.8 | | | HCl | | | 9 |

suspect that the excited state energy is dissipated photochemically in isomerizing the olefinic group. Photochemical olefin isomerization has been observed in $[Ru(bpy)_2(stb)_2]^{2+}$ (stb = 4-styrylpyridine) by Zarnegar et al.[238]. Mixed complexes with phen are listed in Tables 7 and 8 for completeness.

## 4. *Emission from Complexes of Polyazaheterocyclic Ligands*

In this section we include all ruthenium(II) complexes of heterocyclic ligands containing non-coordinating nitrogens whether they are in the heterocyclic ring or not. This grouping brings together a number of otherwise unrelated compounds; in a number of cases the non-ligating nitrogens have been observed to influence excited state properties giving some justification to this organization. Data are tabulated in Tables 9, 10 and 11.

The inductive effect of additional nitrogens in the heterocycle should lead to a stronger ligand pi-acidity. This is well illustrated by the large reduction potentials with ligands **23**[15], **24**[27], **28**[239] and **30**[37] indicating a stabilization of ruthenium(II). However, Crutchley and Lever[36] concluded bipyrazine (bpyz) to be no better a pi-acceptor than bpy; differences between the two are due to a weaker sigma donor strength for bpyz, leaving a higher net charge on ruthenium. In the naphthyridine ligands **31** and **32** presumably a steric effect causes a lowering of the reduction potential below that of $[Ru(bpy)_3]^{2+}$ [30]. The tris chelate of bibenzimidazole (BzImH) demonstrates poor ligand pi-acidity through its reduction potential[28], suggesting poor stabilization of ruthenium(II) by imidazole rings. This is reasonable considering the electron donor properties of N–H groups. On deprotonation one anticipates even less stabilization of ruthenium(II) due to the resultant ligand negative charge and this is indeed realized in $[Ru(bpy)_2(BzIm)]^{+}$ [240].

It is difficult to assess the effect of non-coordinated heteroatoms on emission efficiency because of the lack of quantum yield information. One compound, $[Ru(bpym)_3]^{2+}$ shows an interesting anomaly. Hunziker and Ludi reported an emission quantum yield of 0.09[15] in aqueous solution while Rillema et al.[27] reported 0.011 in propylenecarbonate. The radiative lifetimes estimated from the data of these two reports are very different

**Table 9.** Complexes of polyaza-ligands $[Ru(AB)_3]^{n+}$ and $[Ru(AB)_2(CD)]^{n+}$

| Ligand | Abs | | Em (kK) | | τ (μs) | | Solv. | Φ | | Ref. |
|---|---|---|---|---|---|---|---|---|---|---|
| | kK | E | 77 | RT | 77 | RT | | 77 | RT | |
| (23) | 22.1 | 7400 | | 15.5 | | 0.064 | aq | | 0.09 | 15 |
| | | | | | 7.25x | 0.11 | S | | | 15 |
| (23) | 22.0 | 8600 | | | | | nitrile | | | 27 |
| | | | | 15.6[c] | | 0.131 | pC | | 0.011 | 27 |
| (25) | 19.9 | 10500 | 15.1 | | | | alc | | | 14 |
| | | | | 14.2 | | | $H_2O$ | | | 14 |
| (24) | 22.6 | 15000 | | 16.6 | | 1.04 | $H_2O$ | | | 241 |
| | | | | 16.4[c] | | 0.795 | pC | | 0.074 | 27 |
| | 22.7 | 13000 | | | | | nitrile | | | 27 |
| | 22.8 | | | 15.9 | | 0.72 | nitrile | | 0.024 | 203 |
| | | | | | | 1.1 | DMF | | 0.11 | 203 |
| (26) | 22.1 | 10000 | | 15.8 | | 0.190 | $H_2O$ | | | 14 |
| | | | 17.0 | | | | alc | | | 14 |
| | 22.2 | | | 15.8 | | 0.41 | nitrile | | 0.012 | 203 |
| | | | | | | 0.22 | $H_2O$ | | 0.007 | 203 |
| | | | | | | 0.45 | DMF | | 0.017 | 203 |
| | | | | | | 0.37 | eth | | 0.012 | 203 |
| (27) | 22.5 | 11000 | | 15.7 | | 0.700 | $H_2O$ | | | 14 |
| | | | 16.6 | | | | alc | | | 14 |
| | 22.3 | | | 15.3 | | 1.0 | nitrile | | 0.059 | 203 |
| | | | | | | 0.59 | $H_2O$ | | 0.028 | 203 |
| | | | | | | 0.89 | DMF | | 0.048 | 203 |
| | | | | | | 0.95 | eth | | 0.055 | 203 |
| (28) | 22.9 | 15700 | | 15.0 | | | nitrile | | 0.059 | 239 |
| | | | 16.1 | | 3.1 | | alc | | | 239 |
| (22) | 22.0 | 22000 | | 16.3[c] | | 0.73 | nitrile | | | 9 |
| (30) | 23.0 | 16000 | | | | | nitrile | | | 37 |
| | | | | 17.2 | | 0.200 | $H_2O$ | | | 37 |
| (31) | 19.0 | 12500 | | | | | nitrile | | | 30 |
| | | | 13.6 | | 0.503 | | alc | | | 30 |
| (32) | 17.1 | 9900 | | | | | nitrile | | | 30 |
| | | | 11.9 | | 0.179 | | alc | | | 30 |
| (6) | 20.7 | 9000 | | | | | nitrile | | | 28 |
| | | | 15.4 | | 1.00 | | alc | | | 28 |
| (33) | 26 | 11000 | 13.7 | | 2.0 | | alc | | | 20 |
| $(33)_2$(33-H) | | | 16.6 | | 6 | | alc | | | 20 |
| (33-H) | 24 | 14000 | 16.6 | | 5.9 | | alc | | | 20 |
| (34) | 20.0 | 13500 | | | | | MeCl | | | 3 |
| | | | | 12.8 | | | S | | | 244 |
| (35) | 19.8 | 14000 | | | | | nitrile | | | 4 |
| | | | 12.5 | | 0.01 | | alc | | | 4 |
| $(34)_2$(36) | 19.4 | | | | | | MeCl | | | 3 |
| | | | | 12.9 | | | S | | | 244 |
| (39) | 21.0 | | | 14.9 | | | nitrile | | | 203 |
| | | | | | | 0.11 | $H_2O$ | | | 203 |

**Table 10.** $[RuL_2(bpy)]^{2+}$. (L = polyaza-heterocycle)

| Ligand | Abs | | Em (kK) | | τ (μs) | | Solv. | Φ | | Ref. |
|---|---|---|---|---|---|---|---|---|---|---|
| | kK | E | 77 | RT | 77 | RT | | 77 | RT | |
| (23) | 21.7 | sh | | | | | nitrile | | | 27 |
| | | | | 14.9$^{c}$ | | 0.182 | Pc | | 0.011 | 27 |
| (24) | | | | 15.3$^{c}$ | | 1.099 | Pc | | 0.079 | 27 |
| | 21.6 | 12 000 | | | | | nitrile | | | 27 |
| (28) | 22.7 | 13 200 | | 14.1 | | | nitrile | | | 239 |
| | | | 15.6 | | 2.2 | | alc | | | 239 |
| (34) | 19.3 | 11 600 | | | | | MeCl | | | 3 |
| | | | | 11.7 | | | S | | | 244 |

(ca. 1 vs. 12 μs). Solvent interaction with the uncoordinated nitrogens could be the cause of these differing observations.

In general, protonation of non-ligating nitrogens is reported to quench emission (considered earlier for $[Ru(bpy)_2(CN)_2]$[230]); this is the case for the complexes with bpym[241], bpyz[15] and TAP[37]. In a later study[242] Lever's group found that, after initial quenching by protons, further increase in medium acidity causes emission from $[Ru(bpyz)_3]^{2+}$ to be recovered. The data indicated the presence of several emitting species. They concluded that for the first three protonation steps (to three different bpyz rings) the excited states were stronger bases than the ground states, while the next three protonation steps gave excited states which were weaker bases. As with $[Ru(bpy)_2(CN)_2]$, emission was quenched from this compound by neutral solutions of several other transition elements[243]. No emission enhancement or quenching by silver ion was observed.

**Table 11.** $[RuL(bpy)_2]^{2+}$. (L = polyaza-heterocycle)

| Ligand | Abs | | Em (kK) | | τ (μs) | | Solv. | Φ | | Ref. |
|---|---|---|---|---|---|---|---|---|---|---|
| | kK | E | 77 | RT | 77 | RT | | 77 | RT | |
| (23) | 21.0sh | 11 300 | | 14.7 | | | aq | | | 15 |
| (23) | 20.8 | sh | | | | | nitrile | | | 27 |
| | | | | 14.1$^{c}$ | | 0.076 | pc | | 0.0036 | 27 |
| | 23.8 | 10 500 | | 16.7 | | | nitrile | | | 236 |
| (24) | 21.1 | 9 900 | | | | | nitrile | | | 27 |
| | | | | 14.1$^{c}$ | | 0.376 | pc | | 0.019 | 27 |
| (24) | 21.1 | 9 600 | | 14.5 | | | nitrile | | | 239 |
| | | | 15.6 | | 3.4 | | alc | | | 239 |
| (28) | 22.7 | 14 000 | | 13.5 | | | nitrile | | | 239 |
| | | | 15.1 | | 2.0 | | alc | | | 239 |
| (6) | 21.6 | 12 000 | | | | | nitrile | | | 240 |
| | | | 16.2 | 15.6 | 3.00 | 0.120 | alc | | | 240 |
| (6-H) | 19.7 | 9 100 | | | | | nitrile | | | 240 |
| | | | | 14.2 | | | aq | | | 240 |

$[Ru(BzImH)(bpy)_2]^{2+}$ shows a loss of emission intensity as pH is increased[240]. This could be due to the poorer pi-acidity of the resulting anionic ligand. On the other hand emission from complexes of deprotonated dipyridylamine (ligand **33**) has been observed at low temperature[20]. This ligand is interesting in that $[Ru(Hdpa)_3]^{2+}$ appears to show d-d emission, which is rare in a ruthenium(II) complex.

Complexes of the azo ligands Azpy and NAz have been observed to be only weakly emitting[3, 4, 244]. Resonance raman[244] indicates a pronounced localization of charge on the azo group in the excited state, which could aid energy transfer to the surrounding medium. Also, energy gap considerations suggest poor emission efficiency since the compounds emit at very low energy.

Polyaza ligands have been used to produce oligomeric complexes. Dose and Wilson[236] observed no emission from $[(bpy)_2Ru]_2(bpym)^{4+}$. However, at 15 K Hunziker and Ludi[15] detected an emission from this complex. In contrast, Gafney and co-workers[245] observed emission at room temperature in a fluid medium from a similar compound employing dpp as the bridge.

## 5. *Emission from Complexes of Sulfur-Nitrogen Heterocyclic Ligands*

A few reports have appeared examining ligands containing thiazole rings. Complexes of these aromatic ligands do have MLCT bands and are an interesting departure from the bpy-type. Data for these complexes, and their mixed complexes with bpy, are summarized in Table 12.

The first report of luminescence from a ruthenium(II) complex of this type did not involve an aromatic ligand. Krug and Demas used $[Ru(bt)_3]^{2+}$ [31] to demonstrate that an aromatic system is not essential for CT luminescence. However, there was no emission at room temperature and even at 77 K emission was weak. Efficiency improves in complexes of the related aromatic heterocycle (ligand **38**), which we have observed to be

**Table 12.** $[RuL_n(bpy)_{3-n}]^{2+}$. (L = S,N-heterocycle)

| Ligand | (n) | Abs | | Em (kK) | | τ (μs) | | Solv. | Φ | | Ref. |
|---|---|---|---|---|---|---|---|---|---|---|---|
| | | kK | E | 77 | RT | 77 | RT | | 77 | RT | |
| 37 | (3) | 19.42 | 16600 | 12.7 | | 0.080 | | alc | 0.0003 | | 31 |
| 38 | (3) | 20.6 | | | 13.4 | | | DMSO | | | 246 |
| 38 | (2) | 20.9 | | | 13.4 | | | DMSO | | | 246 |
| 38 | (1) | 21.5 | | | 13.5 | | | DMSO | | | 246 |
| 36 | (3) | 27.0 | | | | | | DMSO | | | 246, 247 |
| 36 | (2) | 22.2 | | | 15.2 | | | DMSO | | | 246 |
| 36 | (1) | 21.8 | | | 15.3 | | | DMSO | | | 246 |
| 39 | (3) | 21.6 | 13500 | | | | | eth | | | 33 |
| | | | | | 16.95 | | | aq | | | 33 |
| 39 | (2) | 21.83 | | | | | | eth | | | 33 |
| | | | | | 16.75 | | | aq | | | 33 |
| 39 | (1) | 22.08 | | | | | | eth | | | 33 |
| | | | | | 16.72 | | | aq | | | 33 |

luminescent at room temperature[246]. Surprisingly, if the other bithiazole isomer (ligand **36**) is coordinated to ruthenium(II) no emission is observed, even at 77 K[247].

We feel the reason for this lack of emission in $[Ru(\mathbf{36})_3]^{2+}$ is low-lying d states. The MLCT transition in this complex appears at 370 nm, considerably higher in energy than the 485 nm of $[Ru(\mathbf{38})_3]^{2+}$ [246]. We would predict the former chelate to be rather photoactive.

Lack of emission from this complex does not appear to be due to any deactivating mode characteristic of the ligand, since the mixed complexes with bpy do indeed emit[246]. Here, the MLCT state lies at much lower energy.

Fitzpatrick and Goodwin have investigated 2-pyridyl-2′-thiazole chelates (ligand **39**)[33]. These were found to be room temperature emitters and very similar to bpy complexes. They noted that the larger "bite" angle of the thiazole ring should lessen sigma-donation, but that the thiazole system is a good π-acceptor. It is probably ligand π-acidity which dominates in the properties of ruthenium(II) complexes.

## *6. Emission from Complexes of Quinoline Derived Ligands*

In 1976 Klassen reported the first ruthenium complex of quinoline derivatives[43]. The steric effect exerted between ligands would be expected to cause weaker binding, less stabilization of the $t_{2g}$ set, and, consequently, a lower energy for the MLCT state. This is indeed realized, with both lower absorption and emission energies in $[Ru(bq)_3]^{2+}$ than in $[Ru(bpy)_3]^{2+}$. Results for complexes of quinoline derived ligands appear in Table 13. When one quinolyl group is replaced by pyridine, the steric effect is lessened and both absorption and emission energies increase somewhat. Although bq has a more extensive pi-system than pq or bpy, the epsilon value for $[Ru(bq)_3]^{2+}$ is less than that for $[Ru(bpy)_3]^{2+}$. This is probably due to a less favorable M–L interaction and results in the room-temperature emission of $[Ru(bq)_3]^{2+}$ being almost completely lost.

A related ligand, DMCH, should be somewhat poorer chelating than bq due to the bridging methylene groups causing non-coplanarity of the quinolines. CV data do show $[Ru(bq)_3]^{2+}$ to be more stabilized than $[Ru(DMCH)_3]^{2+}$ (E for the III/II couples, 1.47 and 1.26 v respectively)[11]. However, $[Ru(DMCH)_3]^{2+}$ has absorption and emission energies almost identical with the bq complex[248, 249]. $[Ru(DMCH)_3]^{2+}$ has a higher epsilon than $[Ru(bq)_3]^{2+}$, but still less than $[Ru(bpy)_3]^{2+}$. The low-emission energy is perhaps the reason for room-temperature emission being of very weak intensity, consistent with energy gap law predictions.

Ruthenium(II) complexes of two ligands, pynapy and dinapy, which should be very similar to pq and DMCH have been investigated[30]. $[Ru(pynapy)_3]^{2+}$ has both absorption and emission bands bathochromically shifted from $[Ru(pq)_3]^{2+}$ by ca. 1600 $cm^{-1}$, which may account for its shorter emission lifetime than the latter complex. $[Ru(dinapy)_3]^{2+}$ also shows a similar shift from $[Ru(DMCH)_3]^{2+}$; however, the shorter lifetime is difficult to compare with this compound, since different solvents were used. The (presumed) non-planarity of dinapy certainly is evident in a comparison of $[Ru(pynapy)_3]^{2+}$ and $[Ru(dinapy)_3]^{2+}$, since the half-wave potentials for the (III/II) couple are consistent with greater ruthenium(II) stabilization in the less distorted compound.

Mixed complexes of bq, pq and DMCH with bpy and phen have also been studied[52, 147–149, 223, 250]. In all cases luminescent states appear to be localized with emis-

**Table 13.** Ruthenium(II) complexes of quinoline derivatives

| Ligands | Abs | | Em (kK) | | τ (μs) | | Solv. | Other | Ref. |
|---|---|---|---|---|---|---|---|---|---|
| | kK | E | 77 | RT | 77 | RT | | | |
| $(42)_3$ | 19.1 | | 13.9 | | 2.9 | | alc | | 43, 52 |
| | 18.8 | 9000 | | 14.2 | | wk | nitrile | | 223 |
| $(43)_3$ | 20.7 | | 15.2 | | 3.8 | | alc | | 43, 52 |
| | | | | 14.1 | | | nitrile | | 147 |
| $(42)_2$bpy | 18.3 | 7250 | 13.5 | | 2.1 | | alc | | 52 |
| | | 8550 | | | | 0.20 | nitrile | | 223 |
| $(42)(bpy)_2$ | 19.0 | 8190 | 13.6 | | 1.9 | | alc | | 52, 250 |
| | | 8710 | 13.7 | 13.5 | 1.4 | 0.27 | nitrile | | 223 |
| $(43)_2$bpy | 20.8 | 11100 | 15.0 | | 3.1 | | nitrile | | 52 |
| $(43)(bpy)_2$ | 21.0sh | | 15.0 | | 3.5 | | nitrile | | 52 |
| | 22.1 | 10500 | | 13.9 | | | | | 147 |
| $(42)_2$phen | 18.1 | 8690 | 13.6 | | 2.3 | | nitrile | | 52 |
| $(42)(phen)_2$ | 19.1 | 9480 | 13.7 | | 2.1 | | nitrile | | 52 |
| $(43)_2$phen | 20.8sh | | 14.8 | | 3.8 | | nitrile | | 52 |
| | 22.1 | 11700 | | | | | | | |
| $(43)(phen)_2$ | 21.0sh | | 15.1 | | 3.8 | | nitrile | | 52 |
| | 22.6 | 13100 | | | | | | | |
| cis[$(42)_2Cl_2$] | 16.1 | 12200 | 11.7 | | 0.46 | | MF | | 248 |
| $(50)_3$ | 18.5 | 10550 | 13.7 | ~13.5 | | wk | nitrile | | 249 |
| | | | 13.6 | | 1.9 | | MF | | 248 |
| $(50)_2$bpy | 17.9 | 9640 | 13.6 | 13.5 | | 0.39 | nitrile | | 249 |
| $(50)(bpy)_2$ | 18.9 | 9410 | 13.9 | 13.6 | | 0.38 | nitrile | | 249 |
| cis[$(50)_2Cl_2$] | 15.7 | 12000 | 11.6 | | | 0.72 | MF | | 249 |
| $(50)(phen)_2$ | | | 14.1 | | 2.0 | | | | 149 |
| $(51)_3$ | | | 18.5 | 18.0 | 96 | | alc | | 251 |
| | | | | | | 0.04 | nitrile | | |
| | 25.1 | 24100 | 18.0 | | 66 | | MF | | 248 |
| | | | | 18.4 | | 0.25 | | | 252 |
| $(51)_2Cl_2$ | 22.9 | 15900 | 13.7 | | 2.2 | | | | 248 |
| $(51)_2$bpy | | | 17.0 | | 5.0 | | alc | | 149 |
| | 22.2sh | 11600 | | 16.3 | | 1.0 | nitrile | At 250k | 252 |
| $(51)(bpy)_2$ | | | 16.9 | | 4.7 | | alc | | 149 |
| | 22.2sh | 11900 | | 16.4 | | 1.1 | nitrile | | 252 |
| $(31)_3$ | 19.0 | 12500 | 13.6 | | 0.503 | | alc | | 30 |
| $(32)_3$ | 17.1 | 9900 | 11.9 | | 0.179 | | alc | | 30 |

sion exclusively from one chromophore. Anderson et al.[147] concluded there to be little L–L coupling in the $[Ru(bq)_n(bpy)_{3-n}]^{2+}$ system. In the mixed complexes with phen, they felt there was evidence for emission from both phen and pq chromophores[148]; but, this was subsequently shown, by Belser et al.[149], to be due to impurities.

One final series of quinoline-based ligand complexes (in which there is no steric effect) has been found to be rather unusual[248, 251, 252]. The ligand i-bq is expected to behave as a substituted bpy. $[Ru(i\text{-}bq)_3]^{2+}$ has a potential (III/II) of 1.12 v indicating less stabilization of ruthenium(II) than in $[Ru(bpy)_3]^{2+}$ [252]. The epsilon value is nearly twice that of $[Ru(bpy)_3]^{2+}$, suggesting that the compound should be an efficient emitter. It turns out that in this complex an entirely different luminescence mechanism is operative. $[Ru(i\text{-}bq)_3]^{2+}$ shows ligand-centered luminescence, which is unusual in a ruthenium complex[251]. This emission is almost an exact copy of that observed from free i-bq with the peaks being bathochromically shifted about 900 $cm^{-1}$. At 77 K the complex luminescence lifetime is 96 μs, which is an order of magnitude greater than that of most ruthenium complexes.

In the mixed complexes, $[Ru(i\text{-}bq)_n(bpy)_{3-n}]^{2+}$, i-bq is not involved in the emission[252]. The absorption spectra show shoulders where $[Ru(bpy)_3]^{2+}$ has its maximum and ε at this point is slightly less than that of this compound. These i-bq mixed complexes do show room temperature-fluid emission, but unfortunately quantum yield values are not available.

Belser et al. have also reported an unusual phenomenon in $[Ru(i\text{-}bq)_2Cl_2]$[248], where metal-centered emission occurs. This is only the second observation of this phenomenon in ruthenium(II) complexes. In addition to $[Ru(i\text{-}bq)_2Cl_2]$, Belser et al. have also observed emission from the cis-dichloro-bis-complexes with bpy, bq, and DMCH[248]. These emissions all occur at low energy as anticipated from the relatively low energy of the first CT band, with low lifetimes. Since most phototubes have poor response in the region of these emissions, it is not unusual that few examples of emission from such compounds have been reported.

## 7. *Emission from Complexes of Tridentate Ligands*

Ruthenium(II) complexes of aromatic tridentate ligands constitute a very small group of complexes. This is understandable, since only a few ligands of this type have been prepared. Nevertheless, much can be learned from observations on this set (data are compiled in Table 14). The prototype of this class of ligands is trpy. One would expect $[Ru(trpy)_2]^{2+}$ to be much like $[Ru(bpy)_3]^{2+}$ in its excited state behavior. However, at room temperature $[Ru(trpy)_2]^{2+}$ is almost non-emitting[19]. The excited state lifetime has been estimated as being between 1 and 5 ns[19, 253] at room temperature. This is quite surprising, since the complex has a larger ε value than $[Ru(bpy)_3]^{2+}$ and at 77 K the emission quantum yield is also greater[188].

Earlier, Crosby's group had developed a multiple state model for $[Ru(bpy)_3]^{2+}$[142, 143, 254], which nicely accounted for the observed temperature quantum yield and lifetime dependence below 77 K, and applied the same model to $[Ru(trpy)_2]^{2+}$[255]. Three states (separated by 55 $cm^{-1}$) accounted for the low-temperature lifetime data, in a picture very similar to that for $[Ru(bpy)_3]^{2+}$[143]. When we calculate the room-temperature lifetime for $[Ru(bpy)_3]^{2+}$ from this model, the proper order of magnitude is obtained, but not so for $[Ru(trpy)_2]^{2+}$. Admittedly this is a long extrapolation, but it suggests that other factors, less important in $[Ru(bpy)_3]^{2+}$, must play a dominant role in dictating the non-radiative decay in $[Ru(trpy)_2]^{2+}$.

Young et al.[253] have observed that $[Ru(trpy)_2]^{2+}$ has a much more "open" structure than does $[Ru(bpy)_3]^{2+}$. Thus, solvents can more readily approach the d orbitals in the former. Van Houten and Watts[174] made this proposition with regard to $[Ru(bpy)_3]^{2+}$ photochemistry and suggested that lowering the solvent polarity (and metal interaction) should enhance the energy of the CT states. Caspar and Meyer[196] have observed a solvent dependence for both lifetime and quantum yield in $[Ru(bpy)_3]^{2+}$; however, it remains to be seen if $[Ru(trpy)_2]^{2+}$ luminescence can be enhanced by proper solvent/matrix choice.

**Table 14.** Complexes of tridentate ligands

| Ligands | Abs | | Em (kK) | | τ (μs) | | Solv. | Φ | | Ref. |
|---|---|---|---|---|---|---|---|---|---|---|
| | kK | E | 77 | RT | 77 | RT | | 77 | RT | |
| $(trpy)_2$ | | | | | 10.7 | | alc | | | 141 |
| | | | | | | | | 0.479 | | 188 |
| | 21.1 | 16200 | | 15.9 | | <5ns | | | | 19 |
| | | | | | | >1.2ns | | | | 253 |
| | | | 16.7 | | | | | | | |
| | | | 15.4 | | 11.0 | | | 0.48 | | 191 |
| | 21.1 | 14600 | | | | | nitrile | | | 239 |
| $(47)_2$ | 17.8 | 8200 | 15.9 | | 12.3 | | | 0.45 | | 191 |
| | | | 14.7 | | | | | | | |
| $(46)_2$ | 20.1 | 38400 | 15.8 | X | 7.8 | ~0.2 | | 0.57 | | 191 |
| | | | 14.5 | | | | | | | |
| $(48)_2$ | 21.0 | 6850 | | | | | $CH_2Cl_2$ | | | 258 |
| $(49)_2$ | 19.5 | 8880 | 14.6 | | 4.77 | | | | | 32 |
| | 16.3sh | 2220 | 14.7 | | 5.6 | | alc | 0.05 | | 191 |
| | 17.8sh | 4800 | 13.4 | | | | | | | 191 |
| $(40)_2$ | 20.7 | | | 14.9 | | | aq | | 0.0012 | 192 |
| $(41)_2$ | 20.0 | 19200 | | 16.5 | | $<5 \times 10^{-3}$ | aq | | | 19 |
| (trpy)(bpy)Cl | 19.9 | 7800 | | 13.7 | | | nitrile | | | 239 |
| | | | 14.4 | | 2.4 | | alc | | | 239 |
| (trpy)(bpy)CN | 20.6 | 10000 | | 15.1 | | | nitrile | | | 239 |
| | | | 16.0 | | 4.6 | | alc | | | 239 |
| (trpy)(bpy)(52) | 22.2 | | | 16.4 | | 0.480 | aq | | $5 \times 10^{-4}$ | 210 |
| (trpy)(bpy)(53) | | | | | | | aq | | $3.8 \times 10^{-4}$ | 210 |

While trpy might be considered simply as an extended form of bpy, this very rigid tridentate ligand may not contain three equal ligating groups. Indeed, it is known that $[Cr(trpy)_2]^{3+}$ deviates significantly from an idealized six-coordinate geometry[256]. Ciantelli et al.[23] observed $[Ru(trpy)_2]^{2+}$ to take up a proton in acid media and Lytle et al. concluded, on the basis of NMR evidence[257] that the terminal nitrogens in this complex could be weakly bonded and undergo photodissociation. The structure of a ruthenium complex containing bidentate trpy has been determined[103], which demonstrates the feasibility of such an argument for weakly bonding trpy.

Thus, it is very reasonable that the other factors involved in the excited state decay of $[Ru(trpy)_2]^{2+}$ include dissociation of at least one pyridyl ligator. Kirchhoff et al.[258] have used an argument based on a kinetic scheme involving photolysis to rationalize inefficient luminescence in $[Ru(trpy)_2]^{2+}$ and related compounds; however, they do not observe extensive photolysis in this system.

A substituted trpy, tro, gives a complex very much like $[Ru(trpy)_2]^{2+}$[191], with a similar quantum yield and lifetime at 77 K. The lowest absorption band is considerably red-shifted from that of $[Ru(trpy)_2]^{2+}$ (3300 $cm^{-1}$), while the $[Ru(tro)_2]^{2+}$ emission is only 800 $cm^{-1}$ below that of the trpy complex. Also, the ε value is very low (8200 $M^{-1}$ $cm^{-1}$). It should be noted that the free ligand, tro, also has a surprisingly weak (ε

5000 $M^{-1}$ $cm^{-1}$) lowest absorption band, which may indicate a unique electronic situation.

While the mono-phenyl ligand, tro, is very much like trpy the tri-phenyl ligand, tsite, is quite different in its ruthenium complex. $[Ru(tsite)_2]^{2+}$ has a very intense lowest absorption band[191], a somewhat higher quantum yield than $[Ru(trpy)_2]^{2+}$ (0.57 vs. 0.48 at 77 K), and is a room temperature emitter (lifetime about 0.2 μs). Unfortunately a room temperature quantum yield has not been reported for comparison with $[Ru(bpy)_3]^{2+}$. Our estimation of the quantum yield at room temperature from the reported lifetime is 0.01, a fraction of that reported for $[Ru(bpy)_3]^{2+}$. Thus, even in $[Ru(tsite)_2]^{2+}$, deactivation modes are more prominent than in $[Ru(bpy)_3]^{2+}$ and the above arguments are most likely applicable.

If phenyl groups are placed on trpy in sterically hindering positions, the effect is dramatic. $[Ru(dpt)_2]^{2+}$ shows no emission even at 77 K[258]. In this complex the lowest absorption band is relatively weak (ε 6850 $M^{-1}$ $cm^{-1}$). Thus, among the trpy ligands there is a crude correlation between ε and quantum yield. It should be noted that energy gap considerations do not indicate a correlation here, since $[Ru(dpt)_2]^{2+}$ and $[Ru(trpy)_2]^{2+}$ absorb at the same energy.

In complexes of the quinoline based ligands we saw room-temperature emission which was weak or absent and could be related to unfavorable steric factors. Most likely $[Ru(dpt)_2]^{2+}$ is non-luminescent for the same reason. Kirchhoff et al.[258] have argued that steric repulsions may cause a $^3MC$ state to lie at lower energy than the $^3CT$ state so that no CT emission occurs. A metal centered state should be more photoactive and $[Ru(dpt)_2]^{2+}$ does indeed undergo photolysis in the presence of nucleophiles. The photolysis product has been formulated as containing a bidentate dpt ligand.

One final tridentate ligand, dqp, has been studied. $[Ru(dqp)_2]^{2+}$ should lie between $[Ru(trpy)_2]^{2+}$ and $[Ru(dpt)_2]^{2+}$ in its properties on the basis of steric arguments. $[Ru(dqp)_2]^{2+}$ was first studied by Klassen et al.[32] and it emits at 77 K. Stone and Crosby also investigated the compound[191] and report a rather low quantum yield (0.05) at 77 K, which they attribute to the sterically hindered structure. No comments appear on photolysis of the complex.

The molar absorbance of this complex is rather low (ε 8880 $M^{-1}$ $cm^{-1}$[32]). This ε value is the same as observed in $[Ru(tro)_2]^{2+}$, but the quantum yields of the two compounds vary by an order of magnitude. We take it that ε comparisons are useful only within a very limited range of ligands.

## *C. Conclusion*

Many authors have commented on the complexity of the emission process. Within a limited series of compounds there is frequently good correlation between emission efficiency and extent of conjugation, molar ε value, or energy gap considerations. However, no predictions can presently be made on going to a new chromophoric system, and this remains an ad-hoc, trial-and-error process.

At present, several factors can be seen as being important in the emission efficiency – molecular structure relationship. The extent of conjugation appears to be one which has been little examined so far. To some degree, the molar epsilon value is a measure of extent of conjugation and appears to have some relationship to efficiency. From energy gap law considerations, it appears that high emission energy is desirable for efficiency, but near-lying d-states must be avoided to minimize nonradiative deactivation through this route. Low-lying d-states are anticipated whenever steric factors weaken the ruthenium-ligand bonding, and such compounds have been found to be only weakly emitting.

In the future, more correlations should be made, which may eventually allow the design of compounds with predicted luminescent properties. Furthermore, we feel the Ferguson model will be widely accepted and may guide subsequent thinking in the design of efficient luminescent ruthenium(II) complexes.

Further correlations of solvent properties with emission efficiency will be valuable, both in helping to understand the basic phenomena and in designing emitting systems for other purposes. Studies on a broad range of complexes based on new ligands will undoubtedly contribute to our understanding of these complicated phenomena. Mixed complexes should be especially useful in answering new questions, such as whether emission always comes from the lowest energy chromophore?

*Acknowledgements.* A point of concern to us is incorrect bibliographic citations; in this paper every citation has been checked against the original article, or in a very few cases, Chemical Abstracts. The author is indebted to Kirsten Krause for this service and for considerable other support during this undertaking. Professor Michael J. Clarke, the editor, is to be thanked for his efforts in polishing the manuscript. Many thanks are also due to Sandy Rivard who patiently typed all of the tables, in spite of my many revisions.

A great many of the authors cited in this paper were kind enough to supply us with collections of their reprints. In this regard thanks are due to Professors F. Bottomley, D. J. Cole-Hamilton, Glenn Crosby, Keith DeArmond, Jim Demas, Harry Goodwin, M.-A. Haga, John Kelly, A. B. P. Lever, A. Ludi, Tom Meyer and A. v. Zelewsky.

A portion of this work was done while the author was a guest in Professor Carl Ballhausen's laboratory. The author is indebted to Carl for his usual hospitality (in spite of an ongoing beer strike at the time!).

# IV. References

1. Sullivan, B. P., Salmon, D. J., Meyer, T. J.: Inorg. Chem. *17,* 3334 (1978)
2. Anderson, S., Seddon, K. R.: J. Chem. Rsch. (S) 74 (1979)
3. Krause, R. A., Krause, K.: Inorg. Chem. *21,* 1714 (1982)
4. Krause, R. A., Krause, K.: ibid. *23,* 2195 (1984)
5. Burstall, F. H.: J. Chem. Soc. 173 (1936)
6. Morgan, G., Burstall, F. H.: ibid. 1649 (1937)
7. Cook, M. J., Lewis, A. P., McAuliffe, G. S. G., Skarda, V., Thomson, A. J., Glasper, J. L., Robbins, D. J.: J. Chem. Soc., Perkin Trans. 2, 1293 (1984)
8. Palmer, R. A., Piper, T. S.: Inorg. Chem. *5,* 864 (1966)
9. Ackermann, M. N., Interrante, L. V.: ibid. *23,* 3904 (1984)
10. Nakamaru, K.: Bull. Chem. Soc. Jpn. *55,* 2697 (1982)
11. Belser, P., von Zelewsky, A.: Helv. Chim. Acta *63,* 1675 (1980)
12. Nasouri, F. G., Blackmore, M. W., Magee, R. J.: Aust. J. Chem. *20,* 1291 (1967)
13. Brown, G. M., Weaver, T. R., Keene, F. R., Meyer, T. J.: Inorg. Chem. *15,* 190 (1976)
14. Kitamura, N., Kawanishi, Y., Tazuke, S.: Chem. Phys. Lett. *97,* 103 (1983)
15. Hunziker, M., Ludi, A.: J. Am. Chem. Soc. *99,* 7370 (1977)
16. Liu, C. F., Liu, N. C., Bailar, J. C., Jr.: Inorg. Chem. *3,* 1085 (1964)
17. Dwyer, F. P., Humpoletz, J. E., Nyholm, R. S.: J. Proc. Roy. Soc. N. S. Wales *80,* 212 (1946)

18. Dwyer, F. P.: ibid. *83,* 134 (1949)
19. Lin, C.-T., Boettcher, W., Chou, M., Creutz, C., Sutin, N.: J. Am. Chem. Soc. *98,* 6536 (1976)
20. Segers, D. P., DeArmond, M. K.: J. Phys. Chem. *86,* 3768 (1982)
21. Steel, P. J., LaHousse, F., Lerner, D., Marzin, C.: Inorg. Chem. *22,* 1488 (1983)
22. Broomhead, J. A., Young, C. G.: Inorg. Syntheses *21,* 127 (1982)
23. Ciantelli, G., Legittimo, P., Pantani, F.: Anal. Chim. Acta *53,* 303 (1971)
24. Goodwin, H. A.: Aust. J. Chem. *17,* 1366 (1964)
25. Alford, P. C., Cook, M. J., Lewis, A. P., McAuliffe, G. S. G., Skarda, V., Thomson, A. J., Glasper, J. L., Robbins, D. J.: J. Chem. Soc., Perkin Trans. *2,* 705 (1985)
26. Dwyer, F. P., Gibson, N. A., Gyarfas, E. C.: J. Proc. Roy. Soc. N. S. Wales *84,* 68 (1950)
27. Rillema, D. P., Allen, G., Meyer, T. J., Conrad, D.: Inorg. Chem. *22,* 1617 (1983)
28. Haga, M.-A.: Inorg. Chim. Acta *77,* L39 (1983)
29. Staniewicz, R. J., Hendricker, D. G.: J. Am. Chem. Soc. *99,* 6581 (1977)
30. Binamira-Soriaga, E., Sprouse, S. D., Watts, R. J., Kaska, W. C.: Inorg. Chim. Acta *84,* 135 (1984)
31. Krug, W. P., Demas, J. N.: J. Am. Chem. Soc. *101,* 4394 (1979)
32. Klassen, D. M., Hudson, C. W., Shaddix, E. L.: Inorg. Chem. *14,* 2733 (1975)
33. Fitzpatrick, L. J., Goodwin, H. A.: Inorg. Chim. Acta *61,* 229 (1982)
34. Braddock, J. N., Meyer, T. J.: J. Am. Chem. Soc. *95,* 3158 (1973)
35. Evans, I. P., Spencer, A., Wilkinson, G.: J. Chem. Soc., Dalton Trans. 204 (1973)
36. Crutchley, R. J., Lever, A. B. P.: Inorg. Chem. *21,* 2276 (1982)
37. Kirsch-De Mesmaeker, A., Nasielski-Hinkens, R., Maetens, D., Pauwels, D., Nasielski, J.: ibid. *23,* 377 (1984)
38. Kirsch-De Mesmaeker, A., Maetens, D., Nasielski-Hinkens, R.: J. Electroanal. Chem. *182,* 123 (1985)
39. Marks, D. N., Siegl, W. O., Gagne, R. R.: Inorg. Chem. *21,* 3140 (1982)
40. Rose, D., Wilkinson, G.: J. Chem. Soc. A 1791 (1970)
41. Gilbert, J. D., Rose, D., Wilkinson, G.: ibid. 2765 (1970)
42. Fabian, R. H., Klassen, D. M., Sonntag, R. W.: Inorg. Chem. *19,* 1977 (1980)
43. Klassen, D. M.: ibid. *15,* 3166 (1976)
44. Anderson, S., Constable, E. C., Seddon, K. R., Turp, J. E., Baggott, J. E., Pilling, M. J.: J. Chem. Soc., Dalton Trans. 2247 (1985)
45. Bottomley, F., Mukaida, M.: ibid. 1933 (1982)
46. Bottomley, F., Mukaida, M.: Inorg. Chim. Acta *97,* L29 (1985)
47. Mercer, E. E., Buckley, R. R.: Inorg. Chem. *4,* 1692 (1965)
48. Broomhead, J. A., Dwyer, F. P., Goodwin, H. A., Kane-Maguire, L., Reid, I.: Inorg. Syntheses *11,* 70 (1968)
49. Dwyer, F. P., Goodwin, H. A., Gyarfas, E. C.: Aust. J. Chem. *16,* 544 (1963)
50. Procedure attributed to H. Goodwin (unpublished work) by: Davies, N. R., Mullins, T. L.: ibid. *20,* 657 (1967)
51. Bryant, G. M., Fergusson, J. E., Powell, H. K. J.: ibid. *24,* 257 (1971)
52. Klassen, D. M.: Chem. Phys. Lett. *93,* 383 (1982)
53. Sprintschnik, G., Sprintschnik, H. W., Kirch, P. P., Whitten, D. G.: J. Am. Chem. Soc. *99,* 4947 (1977)
54. Giordano, P. J., Bock, C. R., Wrighton, M. S.: ibid. *100,* 6960 (1978)
55. McClanahan, S. F., Dallinger, R. F., Holler, F. J., Kincaid, J. R.: ibid. *107,* 4853 (1985)
56. Crutchley, R. J., Lever, A. B. P., Poggi, A.: Inorg. Chem. *22,* 2647 (1983)
57. Birchall, J. D., O'Donoghue, T. D., Wood, J. R.: Inorg. Chim. Acta *37,* L461 (1979)
58. Krause, R. A.: ibid. *31,* 241 (1978)
59. Liu, C. F., Liu, N. C., Bailar, J. C., Jr.: Inorg. Chem. *3,* 1197 (1964)
60. Mahoney, D. F., Beattie, J. K.: ibid. *12,* 2561 (1973)
61. James, B. R., McMillan, R. S., Ochiai, E.: Inorg. Nucl. Chem. Lett. *8,* 239 (1972)
62. Krause, R. A., Krause, K.: Inorg. Chem. *19,* 2600 (1980)
63. Goswami, S., Chakravarty, A. R., Chakravorty, A.: ibid. *20,* 2246 (1981)
64. Weiner, M. A., Basu, A.: ibid. *19,* 2797 (1980)
65. Bressan, M., Rigo, P.: J. Inorg. Nucl. Chem. *38,* 592 (1976)

66. Goodwin, H. A., Lions, F.: J. Am. Chem. Soc. *81,* 311 (1959)
67. Sutton, G. J.: Aust. J. Chem. *19,* 2059 (1966)
68. Johnson, E. C., Sullivan, B. P., Salmon, D. J., Adeyemi, S. A., Meyer, T. J.: Inorg. Chem. *17,* 2211 (1978)
69. Mukaida, M., Yoneda, M., Nomura, T.: Bull. Chem. Soc. Jpn. *50,* 3053 (1977)
70. Tanaka, M., Nagai, T., Miki, E., Mizumachi, K., Ishimori, T.: Nippon Kagaku Kaishi 1112 (1979); Chem. Abstr. *91,* 150494 (1979)
71. Keene, F. R., Salmon, D. J., Meyer, T. J.: J. Am. Chem. Soc. *98,* 1884 (1976)
72. Durham, B., Wilson, S. R., Hodgson, D. J., Meyer, T. J.: ibid. *102,* 600 (1980)
73. Goswami, S., Chakravarty, A. R., Chakravorty, A.: Inorg. Chem. *22,* 602 (1983)
74. Sullivan, B. P., Baumann, J. A., Meyer, T. J., Salmon, D. J., Lehmann, H., Ludi, A.: J. Am. Chem. Soc. *99,* 7368 (1977)
75. Connor, J. A., Meyer, T. J., Sullivan, B. P.: Inorg. Chem. *18,* 1388 (1979)
76. Demas, J. N., Turner, T. F., Crosby, G. A.: ibid. *8,* 674 (1969)
77. Godwin, J. B., Meyer, T. J.: ibid. *10,* 471 (1971)
78. Adeyemi, S. A., Miller, F. J., Meyer, T. J.: ibid. *11,* 994 (1972)
79. Callahan, R. W., Meyer, T. J.: ibid. *16,* 574 (1977)
80. Clear, J. M., Kelly, J. M., O'Connell, C. M., Vos, J. G., Cardin, C. J., Costa, S. R., Edwards, A. J.: J. Chem. Soc., Chem. Commun. 750 (1980)
81. Cole-Hamilton, D. J.: ibid. 1213 (1980)
82. Sullivan, B. P., Caspar, J. V., Johnson, S. R., Meyer, T. J.: Organometallics *3,* 1241 (1984)
83. Choudhury, D., Jones, R. F., Smith, G., Cole-Hamilton, D. J.: J. Chem. Soc., Dalton Trans. 1143 (1982)
84. Kelly, J. M., Vos, J. G.: Angew. Chem. Int. Ed. *21,* 628 (1982)
85. Kelly, J. M., O'Connell, C. M., Vos, J. G.: Inorg. Chim. Acta Lett. *64,* L75 (1982)
86. Red carbonyl solutions were first described by: Chatt, J., Shaw, B. L., Field, A. E.: J. Chem. Soc. 3466 (1964)
87. Black, D. St. C., Deacon, G. B., Thomas, N. C.: Trans. Met. Chem. *5,* 317 (1980)
88. Kelly, J. M., O'Connell, C. M., Vos, J. G.: J. Chem. Soc., Dalton Trans. 253 (1986)
89. Miller, R. R., Brandt, W. W., Puke, M. Sr.: J. Am. Chem. Soc. *77,* 3178 (1955)
90. Morgan, G., Burstall, F. H.: J. Chem. Soc. 1675 (1938)
91. Dwyer, F. P., Goodwin, H. A., Gyarfas, E. C.: Aust. J. Chem. *16,* 42 (1963)
92. James, B. R., McMillan, R. S.: Inorg. Nucl. Chem. Lett. *11,* 837 (1975)
93. Krause, R. A.: Inorg. Chim. Acta *22,* 209 (1977)
94. Alvarez, V. E., Allen, R. J., Matsubara, T., Ford, P. C.: J. Am. Chem. Soc. *96,* 7686 (1974)
95. Ruminski, R. R., Petersen, J. D.: Inorg. Chem. *21,* 3706 (1982)
96. Sullivan, B. P., Calvert, J. M., Meyer, T. J.: ibid. *19,* 1404 (1980)
97. Hudali, H. A., Kingston, J. V., Tayim, H. A.: ibid. *18,* 1391 (1979)
98. Abel, E. W., Bennett, M. A., Wilkinson, G.: J. Chem. Soc. 3178 (1959)
99. Schrock, R. R., Johnson, B. F. G., Lewis, J.: J. Chem. Soc., Dalton Trans. 951 (1974)
100. Robertson, D. R., Robertson, I. W., Stephenson, T. A.: J. Organomet. Chem. *202,* 309 (1980)
101. Irving, R. J.: J. Chem. Soc. 2879 (1956)
102. Kingston, J. V., Jamieson, J. W. S., Wilkinson, G.: J. Inorg. Nucl. Chem. *29,* 133 (1967)
103. Deacon, G. B., Patrick, J. M., Skelton, B. W., Thomas, N. C., White, A. H.: Aust. J. Chem. *37,* 929 (1984)
104. Hiraki, K., Obayashi, Y., Oki, Y.: Bull. Chem. Soc. Jpn. *52,* 1372 (1979)
105. Gopinathan, C.: Inorg. Chim. Acta *87,* L17 (1984)
106. Gilbert, J. D., Wilkinson, G.: J. Chem. Soc. A 1749 (1969)
107. Switkes, E. S., Ruiz-Ramirez, L., Stephenson, T. A., Sinclair, J.: Inorg. Nucl. Chem. Lett. *8,* 593 (1972)
108. Goswami, S., Chakravarty, A. R., Chakravorty, A.: Inorg. Chem. *21,* 2737 (1982)
109. Sullivan, B. P., Salmon, D. J., Meyer, T. J., Peedin, J.: ibid. *18,* 3369 (1979)
110. Vos, J. G., Haasnoot, J. G., Vos, G.: Inorg. Chim. Acta *71,* 155 (1983)
111. Clear, J. M., Kelly, J. M., O'Connell, C. M., Vos, J. G.: J. Chem. Rsch. (M) 3039 (1981)
112. Reveco, P., Medley, J. H., Garber, A. R., Bhacca, N. S., Selbin, J.: Inorg. Chem. *24,* 1096 (1985)

113. Blumer, D. J., Barnett, K. W., Brown, T. L.: J. Organomet. Chem. *173,* 71 (1979)
114. Black, D. St. C., Deacon, G. B., Thomas, N. C.: Inorg. Chim. Acta *54,* L143 (1981)
115. Black, D. St. C., Deacon, G. B., Thomas, N. C.: ibid. *65,* L75 (1982)
116. Black, D. St. C., Deacon, G. B., Thomas, N. C.: Aust. J. Chem. *35,* 2445 (1982)
117. Black, D. St. C., Deacon, G. B., Thomas, N. C.: Polyhedron *2,* 409 (1983)
118. Hough, J. J., Singleton, E.: J. Chem. Soc., Chem. Commun. 371 (1972)
119. Ashworth, T. V., Singleton, E., Hough, J. J.: J. Chem. Soc., Dalton Trans. 1809 (1977)
120. Ashworth, T. V., Reimann, R. H., Singleton, E.: ibid. 1036 (1978)
121. Bernhard, P., Lehmann, H., Ludi, A.: J. Chem. Soc., Chem. Commun. 1216 (1981)
122. Bernhard, P., Burgi, H.-B., Hauser, J., Lehmann, H., Ludi, A.: Inorg. Chem. *21,* 3936 (1982)
123. Bernhard, P., Helm, L., Rapaport, I., Ludi, A., Merbach, A. E.: J. Chem. Soc., Chem. Commun. 302 (1984)
124. Bailey, O. H., Ludi, A.: Inorg. Chem. *24,* 2582 (1985)
125. Bernhard, P., Sargeson, A. M.: J. Chem. Soc., Chem. Commun. 1516 (1985)
126. Dixon, N. E., Lawrance, G. A., Lay, P. A., Sargeson, A. M.: Inorg. Chem. *22,* 846 (1983)
127. Ruiz-Ramirez, L., Stephenson, T. A.: J. Chem. Soc., Dalton Trans. 2244 (1975)
128. Poddar, R. K., Agarwala, U.: J. Inorg. Nucl. Chem. *35,* 567 (1973)
129. Newton, W. E., Searles, J. E.: Inorg. Chim. Acta *7,* 349 (1973)
130. tom Dieck, H., Kollvitz, W., Kleinwachter, I.: Inorg. Chem. *23,* 2685 (1984)
131. Hoggard, P. E., Porter, G. B.: J. Am. Chem. Soc. *100,* 1457 (1978)
132. Wallace, W. M., Hoggard, P. E.: Inorg. Chem. *18,* 2934 (1979)
133. Wallace, W. M., Hoggard, P. E.: ibid. *19,* 2141 (1980)
134. Jones, R. F., Cole-Hamilton, D. J.: Inorg. Chim. Acta Lett. *53,* L3 (1981)
135. Broomhead, J. A., Ileperuma, O.: J. Inorg. Nucl. Chem. *43,* 3181 (1981)
136. Bosnich, B., Dwyer, F. P.: Aust. J. Chem. *19,* 2229 (1966)
137. Durham, B., Walsh, J. L., Carter, C. L., Meyer, T. J.: Inorg. Chem. *19,* 860 (1980)
138. Paris, J. P., Brandt, W. W.: J. Am. Chem. Soc. *81,* 5001 (1959)
139. Kalyanasundaram, K.: Coord. Chem. Rev. *46,* 159 (1982)
140. Klassen, D. M., Crosby, G. A.: J. Chem. Phys. *48,* 1853 (1968)
141. Demas, J. N., Crosby, G. A.: J. Mol. Spectrosc. *26,* 72 (1968)
142. Harrigan, R. W., Crosby, G. A.: J. Chem. Phys. *59,* 3468 (1973)
143. Hager, G. D., Crosby, G. A.: J. Am. Chem. Soc. *97,* 7031 (1975)
144. Crosby, G. A., Hipps, K. W., Elfring, W. H., Jr.: ibid. *96,* 629 (1974)
145. Crosby, G. A., Elfring, W. H., Jr.: J. Phys. Chem. *80,* 2206 (1976)
146. Krause, R. A., Ballhausen, C. J.: Acta Chem. Scand. *A31,* 535 (1977)
147. Anderson, S., Seddon, K. R., Wright, R. D., Cocks, A. T.: Chem. Phys. Lett. *71,* 220 (1980)
148. Cocks, A. T., Wright, R. D., Seddon, K. R.: ibid. *85,* 369 (1982)
149. Belser, P., von Zelewsky, A., Juris, A., Barigelletti, F., Balzani, V.: ibid. *104,* 100 (1984)
150. Sullivan, B. P., Abruna, H., Finklea, H. O., Salmon, D. J., Nagle, J. K., Meyer, T. J., Sprintschnik, H.: ibid. *58,* 389 (1978)
151. Hipps, K. W.: Inorg. Chem. *19,* 1390 (1980)
152. DeArmond, M. K., Carlin, C. M., Huang, W. L.: ibid. *19,* 62 (1980)
153. Carlin, C. M., DeArmond, M. K.: Chem. Phys. Lett. *89,* 297 (1982)
154. Carlin, C. M., DeArmond, M. K.: J. Am. Chem. Soc. *107,* 53 (1985)
155. Felix, F., Ferguson, J., Guedel, H. U., Ludi, A.: ibid. *102,* 4096 (1980)
156. Krol, D. M., Blasse, G.: Chem. Phys. Lett. *77,* 253 (1981)
157. Yersin, H., Gallhuber, E.: J. Am. Chem. Soc. *106,* 6582 (1984)
158. Dallinger, R. F., Woodruff, W. H.: ibid. *101,* 4391 (1979)
159. Angel, S. M., DeArmond, M. K., Donohoe, R. J., Hanck, K. W., Wertz, D. W.: ibid. *106,* 3688 (1984)
160. Caspar, J. V., Westmoreland, T. D., Allen, G. H., Bradley, P. G., Meyer, T. J., Woodruff, W. H.: ibid. *106,* 3492 (1984)
161. Krausz, E.: Chem. Phys. Lett. *116,* 501 (1985)
162. Ferguson, J., Herren, F.: ibid. *89,* 371 (1982)
163. Ferguson, J., Krausz, E. R.: ibid. *93,* 21 (1982)
164. Ferguson, J., Krausz, E. R., Maeder, M.: J. Phys. Chem. *89,* 1852 (1985)

165. Ferguson, J., Herren, F., Krausz, E. R., Maeder, M., Vrbancich, J.: Coord. Chem. Rev. *64,* 21 (1985)
166. Gafni, A., Steinberg, I. Z.: Isr. J. Chem. *15,* 102 (1977)
167. Mandal, K., Demas, J. N.: Chem. Phys. Lett. *84,* 410 (1981)
168. Mandal, K., Pearson, T. D. L., Demas, J. N.: J. Chem. Phys. *73,* 2507 (1980)
169. Kirk, A. D., Hoggard, P. E., Porter, G. B., Rockley, M. G., Windsor, M. W.: Chem. Phys. Lett. *37,* 199 (1976)
170. Kober, E. M., Meyer, T. J.: Inorg. Chem. *21,* 3967 (1982)
171. Kober, E. M., Meyer, T. J.: ibid. *23,* 3877 (1984)
172. Kober, E. M., Sullivan, B. P., Meyer, T. J.: ibid. *23,* 2098 (1984)
173. Demas, J. N., Adamson, A. W.: J. Am. Chem. Soc. *95,* 5159 (1973)
174. Van Houten, J., Watts, R. J.: ibid. *98,* 4853 (1976)
175. Van Houten, J., Watts, R. J.: Inorg. Chem. *17,* 3381 (1978)
176. Allsopp, S. R., Cox, A., Kemp, T. J., Reed, W. J.: J. Chem. Soc., Faraday Trans. 1, *74,* 1275 (1978)
177. Gleria, M., Minto, F., Beggiato, G., Bortolus, P.: J. Chem. Soc., Chem. Commun. 285 (1978)
178. Porter, G. B., Sparks, R. H.: J. Photochem. *13,* 123 (1980)
179. Durham, B., Caspar, J. V., Nagle, J. K., Meyer, T. J.: J. Am. Chem. Soc. *104,* 4803 (1982)
180. Henderson, L. J., Jr., Fronczek, F. R., Cherry, W. R.: ibid. *106,* 5876 (1984)
181. Reveco, P., Schmehl, R. H., Cherry, W. R., Fronczek, F. R., Selbin, J.: Inorg. Chem. *24,* 4078 (1985)
182. Allen, G. H., White, R. P., Rillema, D. P., Meyer, T. J.: J. Am. Chem. Soc. *106,* 2613 (1984)
183. Caspar, J. V., Kober, E. M., Sullivan, B. P., Meyer, T. J.: ibid. *104,* 630 (1982)
184. Caspar, J. V., Meyer, T. J.: J. Phys. Chem. *87,* 952 (1983)
185. Caspar, J. V., Meyer, T. J.: Inorg. Chem. *22,* 2444 (1983)
186. Kober, E. M., Meyer, T. J.: ibid. *24,* 106 (1985)
187. Calvert, J. G., Pitts, J. N.: *Photochemistry;* John Wiley: New York, 1966; p. 174
188. Demas, J. N., Crosby, G. A.: J. Am. Chem. Soc. *93,* 2841 (1971)
189. Watts, R. J., Crosby, G. A.: ibid. *93,* 3184 (1971)
190. Cook, M. J., Lewis, A. P., McAuliffe, G. S. G., Skarda, V., Thomson, A. J., Glasper, J. L., Robbins, D. J.: J. Chem. Soc., Perkin Trans. 2, 1303 (1984)
191. Stone, M. L., Crosby, G. A.: Chem. Phys. Lett. *79,* 169 (1981)
192. Nakamaru, K., Nishio, K., Nobe, H.: Sci, Rep. Hirosaki Univ. *26,* 57 (1979)
193. Bolletta, F., Juris, A., Maestri, M., Sandrini, D.: Inorg. Chim. Acta *44,* L175 (1980)
194. Van Houten, J., Watts, R. J.: J. Am. Chem. Soc. *97,* 3843 (1975)
195. Caspar, J. V., Sullivan, B. P., Kober, E. M., Meyer, T. J.: Chem. Phys. Lett. *91,* 91 (1982)
196. Caspar, J. V., Meyer, T. J.: J. Am. Chem. Soc. *105,* 5583 (1983)
197. Bolletta, F., Maestri, M., Moggi, L., Balzani, V.: J. Phys. Chem. *78,* 1374 (1974)
198. Vining, W. J., Caspar, J. V., Meyer, T. J.: ibid. *89,* 1095 (1985)
199. Yersin, H., Hensler, G., Gallhuber, E., Rettig, W., Schwan, L. O.: Inorg. Chim. Acta *105,* 201 (1985)
200. Kurimura, Y., Yokota, H., Shigehara, K., Tsuchida, E.: Bull. Chem. Soc. Jpn. *55,* 55 (1982)
201. Nakamaru, K.: ibid. *55,* 1639 (1982)
202. Gutmann, V.: Electrochim. Acta *21,* 661 (1976)
203. Kawanishi, Y., Kitamura, N., Kim, Y., Tazuke, S.: Sci. Pap. Inst. Phys. Chem. Res. (Jpn.) *78,* 212 (1984)
204. Belser, P., von Zelewsky, A., Juris, A., Barigelletti, F., Balzani, V.: Gazz. Chim. Ital. *115,* 723 (1985)
205. Wolfgang, S., Gafney, H. D.: J. Phys. Chem. *87,* 5395 (1983)
206. Wei, S., Gafney, H. D., Clark, J. B., Perettie, D. J.: Chem. Phys. Lett. *99,* 253 (1983)
207. DeWilde, W., Peeters, G., Lunsford, J. H.: J. Phys. Chem. *84,* 2306 (1980)
208. Demas, J. N., Addington, J. W.: J. Am. Chem. Soc. *98,* 5800 (1976)
209. Haga, M.-A., Kanai, H., Tanaka, T.: Bull. Fac. Ed., Mie Univ. *34,* 59 (1983)
210. Root, M. J., Deutsch, E., Sullivan, J. C., Meisel, D.: Chem. Phys. Lett. *101,* 353 (1983)
211. Mingardi, M., Porter, G. B.: Spectrosc. Lett. *1,* 293 (1968)
212. Zuloaga, F., Jauregui, G.: Rev. Latinoamer. Quim. *2,* 97 (1971); Chem. Abstr. *76,* 52290 (1972)

213. Vogler, A., Kunkely, H.: Inorg. Chim. Acta Lett. *53,* L215 (1981)
214. Crosby, G. A., Hager, G. D., Hipps, K. W., Stone, M. L.: Chem. Phys. Lett. *28,* 497 (1974)
215. Wrighton, M. S., Pdungsap, L., Morse, D. L.: J. Phys. Chem. *79,* 66 (1975)
216. Klassen, D. M., Crosby, G. A.: J. Mol. Spectrosc. *25,* 398 (1968)
217. Arnold, G. S., Klotz, W. L., Halper, W., DeArmond, M. K.: Chem. Phys. Lett. *19,* 546 (1973)
218. Antipas, A., Buchler, J. W., Gouterman, M., Smith, P. D.: J. Am. Chem. Soc. *100,* 3015 (1978)
219. Rillema, D. P., Nagle, J. K., Barringer, L. F., Jr., Meyer, T. J.: ibid. *103,* 56 (1981)
220. Demas, J. N., Crosby, G. A.: J. Phys. Chem. *75,* 991 (1971)
221. Elfring, W. H., Jr., Crosby, G. A.: J. Am. Chem. Soc. *103,* 2683 (1981)
222. Basu, A., Weiner, M. A., Strekas, T. C., Gafney, H. D.: Inorg. Chem. *21,* 1085 (1982)
223. Juris, A., Barigelletti, F., Balzani, V., Belser, P., von Zelewsky, A.: Isr. J. Chem. *22,* 87 (1982)
224. Baggott, J. E., Gregory, G. K., Pilling, M. J., Anderson, S., Seddon, K. R., Turp, J. E.: J. Chem. Soc., Faraday Trans. 2, *79,* 195 (1983)
225. Ford, W. E., Calvin, M.: Chem. Phys. Lett. *76,* 105 (1980)
226. Ferguson, J., Mau, A. W.-H., Sasse, W. H. F.: ibid. *68,* 21 (1979)
227. Johansen, O., Kowala, C., Mau, A. W.-H., Sasse, W. H. F.: Aust. J. Chem. *32,* 2395 (1979)
228. Demas, J. N., Addington, J. W., Peterson, S. H., Harris, E. W.: J. Phys. Chem. *81,* 1039 (1977)
229. Bartocci, C., Bignozzi, C. A., Scandola, F., Rumin, R., Courtot, P.: Inorg. Chim. Acta *76,* L119 (1983)
230. Peterson, S. H., Demas, J. N.: J. Am. Chem. Soc. *98,* 7880 (1976)
231. Peterson, S. H., Demas, J. N.: ibid. *101,* 6571 (1979)
232. Kinnaird, M. G., Whitten, D. G.: Chem. Phys. Lett. *88,* 275 (1982)
233. Bignozzi, C. A., Scandola, F.: Inorg. Chem. *23,* 1540 (1984)
234. Bignozzi, C. A., Scandola, F.: Inorg. Chim. Acta *86,* 133 (1984)
235. Shriver, D. F., Posner, J.: J. Am. Chem. Soc. *88,* 1672 (1966)
236. Dose, E. V., Wilson, L. J.: Inorg. Chem. *17,* 2660 (1978)
237. Wacholtz, W. M., Auerbach, R. A., Schmehl, R. H., Ollino, M., Cherry, W. R.: ibid. *24,* 1758 (1985)
238. Zarnegar, P. P., Bock, C. R., Whitten, D. G.: J. Am. Chem. Soc. *95,* 4367 (1973)
239. Balzani, V., Juris, A., Barigelletti, F., Belser, P., von Zelewsky, A.: Sci. Pap. Inst. Phys. Chem. Res. (Jpn.) *78,* 78 (1984)
240. Haga, M.-A.: Inorg. Chim. Acta *75,* 29 (1983)
241. Crutchley, R. J., Lever, A. B. P.: J. Am. Chem. Soc. *102,* 7128 (1980)
242. Crutchley, R. J., Kress, N., Lever, A. B. P.: ibid. *105,* 1170 (1983)
243. Haga, M.-A., Dodsworth, E. S., Eryavec, G., Seymour, P., Lever, A. B. P.: Inorg. Chem. *24,* 1901 (1985)
244. Wolfgang, S., Strekas, T. C., Gafney, H. D., Krause, R. A., Krause, K.: ibid. *23,* 2650 (1984)
245. Braunstein, C. H., Baker, A. D., Strekas, T. C., Gafney, H. D.: ibid. *23,* 857 (1984)
246. Krause, R. A., Krause, K.: Work in progress
247. Krug, W. P., Demas, J. N.: presented at the 178th National Meeting of the American Chemical Society, Washington, D.C., Sept. 1979. INOR 99
248. Belser, P., von Zelewsky, A., Juris, A., Barigelletti, F., Balzani, V.: Gazz. Chim. Ital. *113,* 731 (1983)
249. Juris, A., Balzani, V., Belser, P., von Zelewsky, A.: Helv. Chim. Acta *64,* 2175 (1981)
250. Barigelletti, F., Juris, A., Balzani, V., Belser, P., von Zelewsky, A.: Inorg. Chem. *22,* 3335 (1983)
251. Belser, P., von Zelewsky, A., Juris, A., Barigelletti, F., Tucci, A., Balzani, V.: Chem. Phys. Lett. *89,* 101 (1982)
252. Juris, A., Barigelletti, F., Balzani, V., Belser, P., von Zelewsky, A.: Inorg. Chem. *24,* 202 (1985)
253. Young, R. C., Nagle, J. K., Meyer, T. J., Whitten, D. G.: J. Am. Chem. Soc. *100,* 4773 (1978)
254. Harrigan, R. W., Hager, G. D., Crosby, G. A.: Chem. Phys. Lett. *21,* 487 (1973)
255. Agnew, S. F., Stone, M. L., Crosby, G. A.: ibid. *85,* 57 (1982)

256. Wickramasinghe, W. A., Bird, P. H., Jamieson, M. A., Serpone, N.: J. Chem. Soc., Chem. Commun. 798 (1979)
257. Lytle, F. E., Petrosky, L. M., Carlson, L. R.: Anal. Chim. Acta *57,* 239 (1971)
258. Kirchhoff, J. R., McMillin, D. R., Marnot, P. A., Sauvage, J.-P.: J. Am. Chem. Soc. *107,* 1138 (1985)
259. Harriman, A.: J. Chem. Soc., Chem. Commun. 777 (1977)
260. Lytle, F. E., Hercules, D. M.: J. Am. Chem. Soc. *91,* 253 (1969)
261. Mandal, K., Pearson, T. D. L., Demas, J. N.: Inorg. Chem. *20,* 786 (1981)
262. Johansen, O., Kowala, C., Mau, A. W.-H., Sasse, W. H. F.: Aust. J. Chem. *32,* 1453 (1979)
263. DeLaive, P. J., Lee, J. T., Sprintschnik, H. W., Abruna, H., Meyer, T. J., Whitten, D. G.: J. Am. Chem. Soc. *99,* 7094 (1977)
264. Belser, P., von Zelewsky, A., Zehnder, M.: Inorg. Chem. *20,* 3098 (1981)
265. Shimidzu, T., Izaki, K., Akai, Y., Iyoda, T.: Polym. J. (Tokyo) *13,* 889 (1981)
266. Calvert, J. M., Caspar, J. V., Binstead, R. A., Westmoreland, T. D., Meyer, T. J.: J. Am. Chem. Soc. *104,* 6620 (1982)
267. Rillema, D. P., Callahan, R. W., Mack, K. B.: Inorg. Chem. *21,* 2589 (1982)
268. Demas, J. N., Harris, E. W., McBride, R. P.: J. Am. Chem. Soc. *99,* 3547 (1977)
269. Lay, P. A., Sasse, W. H. F.: Inorg. Chem. *23,* 4123 (1984)
270. Kelly, J. M., Long, C., O'Connell, C. M., Vos, J. G., Tinnemans, A. H. A.: ibid. *22,* 2818 (1983)
271. Crosby, G. A., Perkins, W. G., Klassen, D. M.: J. Chem. Phys. *43,* 1498 (1965)
272. Watts, R. J., Crosby, G. A.: J. Am. Chem. Soc. *94,* 2606 (1972)
273. Hauenstein, B. L., Jr., Dressick, W. J., Buell, S. L., Demas, J. N., DeGraff, B. A.: ibid. *105,* 4251 (1983)
274. Wallace, W. L., Bard, A. J.: J. Phys. Chem. *83,* 1350 (1979)

# Platinum Amine Coordination Compounds as Anti-Tumor Drugs. Molecular Aspects of the Mechanism of Action

**Jan Reedijk[1], Anne Marie J. Fichtinger-Schepman[2], Allan T. van Oosterom[3] and Piet van de Putte[4]**

1 Department of Chemistry, Gorlaeus Laboratories, Leiden University, P.O. Box 9502, 2300 RA Leiden, The Netherlands
2 Medical Biological Laboratory, TNO, 2280 AA Rijswijk, The Netherlands
3 Department of Clinical Oncology, Leiden University, The Netherlands
4 Biochemical Laboratory, Leiden University, The Netherlands

An overview is presented of platinum amine compounds, used as antitumor drugs. The history of the discovery is briefly described, including a discussion of the synthesis of platinum amine coordination compounds. Attention is given to the relationship between biological activity and structure of the compounds, especially to the nature of the ligands coordinated to platinum. The primary biological event appears to be the interaction with nucleic acids, and therefore the major part of the review is deals with studies of the binding interactions of platinum amine compounds (active and inactive ones) to nucleic acids and nucleic acid fragments. Studies on mononucleic acid fragments have made clear that a strong preference exists for binding at guanine-N7 sites. Studies on oligonucleotides have shown that, when two neighboring guanines are present, chelation of the *cis*-$Pt(NH_3)_2$ unit is strongly favored above all other possibilities. However, binding at neighboring AG-units also occurs readily. Studies on DNA (in vivo and in vitro) have shown that similar bindings occur in DNA and oligonucleotides, and that the DNA is distorted in a similar manner to double-stranded oligonucleotides.



Structure and Bonding 67

# Table of Abbreviations

| | |
|---|---|
| Cisplatin | *cis*-$PtCl_2(NH_3)_2$ |
| CHIP | $PtCl_2(OH)_2$(isopropylamine)$_2$ |
| CBDCA | *cis*-$Pt(C_6H_6O_4)(NH_3)_2$ |
| NMR | nuclear magnetic resonance |
| FPLC | fast protein liquid chromatography |
| HPLC | high performance liquid chromatography |
| G | guanine and guanosine |
| GMP | guanosine monophosphate |
| C | cytidine and cytosine |
| A | adenine and adenosine |
| T | thymine and thymidine |
| B | any nucleobase |
| DNA | deoxyribonucleic acid |
| ss | single stranded |
| ds | double stranded |
| NOE | nuclear Overhauser effect |
| AAS | atomic absorption spectroscopy |
| Am | amine ligand |
| JM-40 | Pt(en) (malonate) |
| en | ethylenediamine |
| tmen | N,N,N′,N′-tetramethyl ethylene diamine |
| dmdap | 2,2-dimethyl-1,2-diaminopropane |
| p | (terminal) phosphate group |
| d | deoxyribose |

# 1 Introduction

Platinum amine coordination compounds have been known for more than a century and have fascinated chemists from the very beginning[1,2]. Through the work of Alfred Werner[3] it became understood that a compound such as $PtCl_2(NH_3)_2$ – which was first described by Peyrone[4] in 1845 – could exist in two geometrical isomers (*cis* and *trans*) that are kinetically stable enough to be isolated from solution and kept in the solid state. Many other Pt(II) compounds with amines as ligands have been reported since then, as will be described below. Platinum(IV) compounds with amine ligands were also known from the early days, and these appeared to be even more kinetically inert than the corresponding Pt(II) compounds[5]. Some examples of Pt(II) and Pt(IV) amine coordination compounds are shown in Fig. 1.

In the early days of coordination chemistry the platinum compounds were mainly of interest for the mechanically-oriented inorganic chemists, and research on such compounds after the second world war has provided the basis for modern mechanistic inorganic chemistry[5–8]. Especially the process of ligand exchange and the kinetics for different ligands and different geometries, were understood much better as a result of the wealth of data from platinum compounds. The fact that square-planar platinum(II) compounds do occur in cis- and trans-isomers that are slowly – if at all – interconverted, has allowed kinetic studies for both isomers of the same species. The reaction velocity of ligand substitution is largely determined by the nature of the ligand trans to the leaving ligand. Without the availability of a large number of Pt compounds the development of this empirical, but very useful rule for trans substitution would have been impossible. The application of this trans-effect rule in the synthesis of Pt(II) compounds will be described in the next section.

More recently, it became apparent that planar Pt(II) compounds are also of interest from a physics point of view, since – partly oxidized – square-planar coordinated Pt(II) compounds frequently form stacked species in the solid state, giving rise to one-dimensional conduction[9].

In the mid-sixties, it became clear that also from a biological point of view these compounds are very interesting. Rosenberg and his group[10] initiated this renewed interest by studying the growth of *E. coli* bacteria under the influence of an electric field. This field was generated between two „inert“ platinum electrodes and aqueous $NH_4Cl$ was used as an electrolyte. The bacteria showed a strong filamenteous growth, which was initially not understood[11]. However, subsequent experiments by the same group[12] soon made clear that the filamenteous growth was not caused by the electric field, but by the presence of small amounts of Pt(IV) compounds in the solution. These compounds were, apparently, formed from the Pt-electrodes and the corroding $NH_4Cl$ solution under the influence of the electric field. Detailed investigations showed that the compounds present in solution were, among others, *cis*-$PtCl_4(NH_3)_2$. In subsequent microbiological studies the *cis*-Pt(II) species – now called cisplatin – turned out to be the most active in causing filament formation.

After that, tests were undertaken to study the growth-reducing capacity of cisplatin on animal tumors, such as Sarcoma 180 in Swiss white mice[13]. The results were so positive – in many cases total regression of the tumors was observed – that clinical trials were soon thereafter planned and performed. The first studies, in 1972, demonstrated remarkable anti tumor activity[14] for cisplatin. However, large-scale applications had to

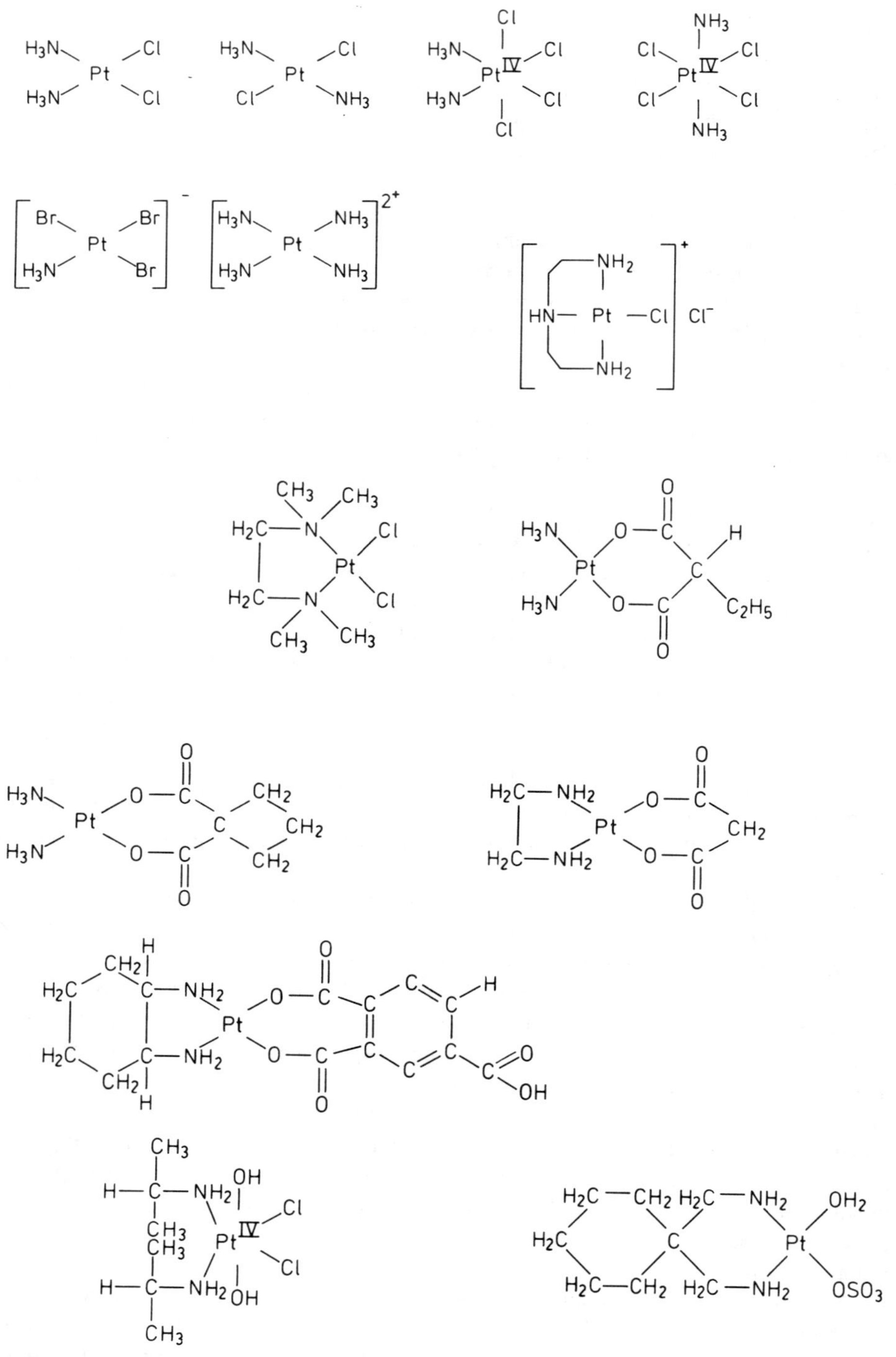

**Fig. 1.** Examples of Pt(II) and Pt(IV) amine coordination compounds; most of these are playing a role in clinical or mechanistic investigations

wait for some more years, since toxic side effects were severe. To be mentioned are: nausea, vomiting, bone marrow toxicity, neurotoxicity and renal toxicity[14–16].

Since the late seventies, the clinical application of cisplatin has increased enormously[16–18], mainly as a result of improved administration procedures and its use in combination therapy, i.e. the simultaneous application of a variety of synergistic anti-tumor drugs. Nowadays, usually a dosage per patient of about 100 mg of cisplatin per $m^2$ body surface area, dissolved in saline, is given intravenously every month by standard administration protocols.

Final FDA approval in the USA in 1979, has led to a further increasing number of applications of cisplatin and as a result it has become the leading and most widely used anticancer drug (30,000 patients cured each year in the USA). The drug is also registered widely in Europe and in Japan, and its use is still increasing. Complete remissions for testicular cancers in more than 85% of all treated patients are obtained, whereas high effectivity in the treatment of ovarian and bladder cancer and considerable activity in osteogenic sarcoma, head and neck cancer, endometrial and cervical cancer and non-small cell lung cancer have been reported.

It should be noted that the corresponding trans isomer of cisplatin exhibits no anti-tumor activity.

Investigations on the working mechanism of cisplatin have been carried out during the last decade by a variety of research groups, involving chemists, biochemists, biologists and medical researchers[17]. Early studies already made it clear that reaction of platinum compounds with nucleic acids play an important role in the mechanism of action. The present review deals with the status of this field, with special attention to platinum-DNA interactions.

It should be mentioned here that several very good reviews have appeared during the last decade, describing a wide variety of aspects of platinum-anti-tumor compounds[14–21, 146] and of binding of metal compounds to nucleic acids and fragments[22–25, 162, 163]. The present review will give major attention to the molecular aspects of the mechanism, starting with structure-activity correlations, followed by the detailed changes of the DNA conformation induced by binding of the Pt compounds to it, and ending with a description of the in vivo formed products upon cisplatin treatment.

## 2 Synthesis and Structure of Platinum Amine Coordination Compounds

As mentioned in the introduction, many platinum amine compounds have been reported since the beginning of the century, and some compounds have been reported even as early as 1845[1, 3, 4]. The chemistry of platinum in aqueous systems largely involves oxidation states Pt(II) and Pt(IV). The coordination geometries of these two states are, almost invariably, square-planar and octahedral, respectively, as is seen from the structures drawn in Fig. 1. Common starting reagents for the chemistry of platinum are $K_2PtCl_4$ and $K_2PtCl_6$. Starting with $K_2PtCl_4$, treatment with amines (e.g. $NH_3$) in aqueous solution may yield the following compounds:

- $[Pt(NH_3)_4][PtCl_4]$ (the so-called Magnus salt; poorly soluble)
- $\alpha$-$PtCl_2(NH_3)_2$ (the cis-isomer; formerly called Peyrone's chloride)
- $\beta$-$PtCl_2(NH_3)_2$ (the trans-isomer; formerly called Reiset's chloride).

With an excess of amine ligands also $[Pt(amine)_4]Cl_2$ may be obtained, whereas also mixed species, such as $K[PtCl_3(NH_3)]$ and $[PtCl(NH_3)_3]Cl$ are known. For a full discussion of these compounds the reader is referred to the earlier literature[1,3,26].

Since the early investigations[26] learned that the amine platinum compounds with anti-tumor properties all have the cis-geometry and can be represented by the general formulae: $PtX_2(ligand)_2$ and $PtX_2Y_2(ligand)_2$, we will concentrate on the synthesis of such species. Therefore some more information about the above-mentioned trans-effect is required. This effect is most easily explained starting from the species $PtX_3L^-$ (in fact the first reaction product formed from $PtX_4^{2-}$ and ligand L) in the reaction with another ligand molecule, called N. The reaction product may then have either of the following structures:

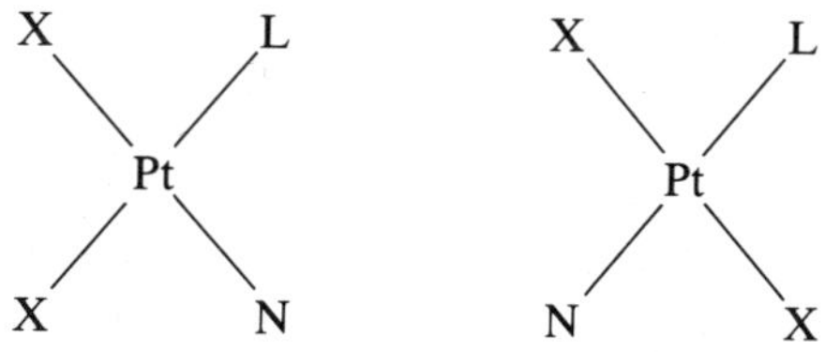


cis trans

The relative portion of both isomers is determined by the nature of the leaving group in $PtX_3L^-$ (i.e. X or L). Ligands differ in their capabilities of directing an incoming ligand (N in this case) to a position trans to themselves. Careful studies in the past have resulted in a sequence of trans-directing power for several kinds of ligands. The strongest trans effects are exhibited by ligands such as $CN^-$, CO; somewhat weaker effects occur in the sequence:

$I^- > Br^- > Cl^- > NH_3$, whereas $OH^-$ and $H_2O$ have very weak trans effects[7].

So, starting from $PtCl_4^{2-}$ reaction with $NH_3$ should, consecutively, produce $PtCl_3(NH_3)^-$ and *cis*-$PtCl_2(NH_3)_2$. In practice, however, often also some trans-isomer is found, because Cl and $NH_3$ are close in the trans-labilizing series. Therefore, many syntheses are carried out with $PtI_4^{2-}$ as the starting product (easily obtained from $K_2PtCl_4$ and an excess of iodide[27]). In performing this reaction, one should realize that the substitution reactions are not only determined by the trans effect. In fact, treatment of *cis*-$PtI_2(amine)_2$ with $Cl^-$ would result in removal of the ligands trans to $I^-$, i.e. the amine, which is undesired. The thermodynamic strength of the Pt-amine bond in this case retards or even prevents the reaction, and in fact the $I^-$ would be released – slowly – with an excess of chloride. This reaction is not very efficient, however, and therefore a faster method has been used to prepare the *cis*-$PtCl_2(amine)_2$ from the iodide[28]. When using two equivalents of $AgNO_3$, a precipitate of AgI is formed in aqueous solution, leaving the cationic species $[Pt(amine)_2(H_2O)_2]^{2+}$ in solution, which now can easily be converted

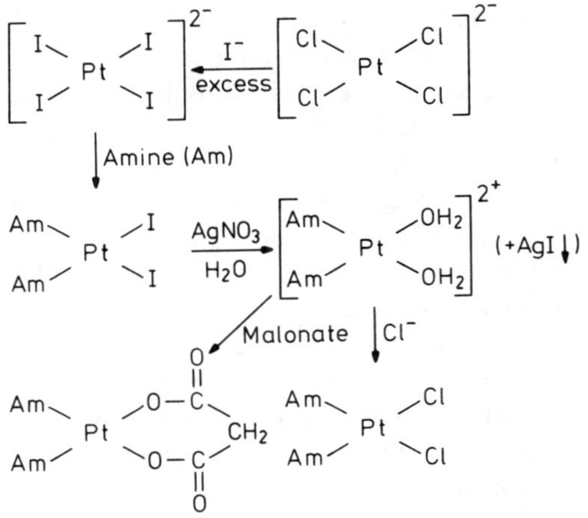


**Fig. 2.** Synthetic methods for Pt(II) amine compounds

with $Cl^-$ to *cis*-$PtCl_2(amine)_2$ (amines result in strong Pt–N bonds and have a stronger trans-effect than water, vide supra) (see Fig. 2). Other compounds, e.g. the malonate can be prepared similarly. In case of chelating didentate amine ligands, such as ethylene diamine (en), direct reaction with $PtCl_4^{2-}$ results in $PtCl_2$(diamine). Nevertheless, even in these cases the synthesis through the iodide is often used[28].

Platinum(IV) amine compounds can be prepared using similar procedures, starting from $PtCl_6^{2-}$. Other preparations[1] use the Pt(II) amine compounds together with an oxidizing agent, such as $H_2O_2$, $I_2$ or $Cl_2$. Some reactions are given schematically in Fig. 3.

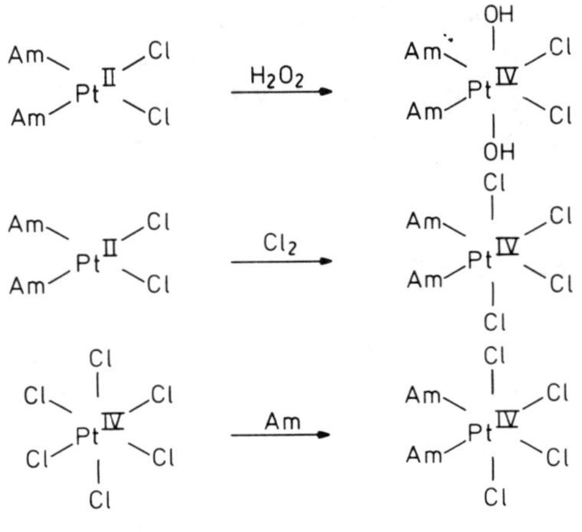


**Fig. 3.** Synthetic methods for Pt(IV) amine compounds

## 3 Relationships Between Structure and Activity

Soon after the first reports about the biological activity of the classical compound *cis*-$PtCl_2(NH_3)_2$, an extensive investigation was started in a search for analogs with similar activities and – hopefully – improved properties as a drug. The first extensive results of these studies were reported[26, 29] at a special symposium, held in Dallas (1977). Early studies by many others have resulted in a large group of analogs with similar biological activities; some of these are used in extensive clinical trials in many countries and hospitals[13, 17, 30]. Recent evidence seems to indicate that some of these analogs have superior properties in the treatment of tumors in patients who do not respond to cisplatin[30]. The most promising compounds, considered today, are redrawn in Fig. 4.

The studies with a large number of Pt(II) and Pt(IV) compounds prepared by others and us during the last decade, have made clear that Pt compounds with anti-tumor activity have to fulfil all of the following structural requirements:

1. The two amine ligands in the Pt-compound should be in a cis-orientation. In a didentate chelating ligand, this geometric requirement is automatically fulfilled. The general formulae should be *cis*-$PtX_2(Am)_2$ and *cis*-$PtY_2X_2(Am)_2$. It should be noted here, however, that the number of variations studied in the case of Pt(IV) is still rather limited.
2. The ligands X, usually anions, should consist of groups that have intermediate binding strength to Pt(II), or that are – for other reasons – easily leaving (i.e. by enzymatic activation). Examples are: $Cl^-$, $SO_4^{2-}$; citrate(3−), oxalate(2−) and other carboxylic acid residues.
   For the Pt(IV) compounds, the Y group is often $OH^-$ (with the two Y ligands in trans-orientation); only a few anti-tumor active compounds of this class have been reported[26, 31, 32].
3. The amine ligands, either monodentate or didentate, should have at least one N–H group, i.e. possess a hydrogen-bond donor function[31]. All compounds with both amine ligands lacking such a H-bond donor property, have been found to be inactive. The role of this N–H group in the biological activity, however, is far from being understood. It could be either kinetic (i.e. play a role in the approach of the DNA, vide infra), or thermodynamic (e.g. give an additional (de)stabilization after binding to the biological target DNA; vide infra). However, also steric effects and a role in transport through the cell wall cannot be excluded.

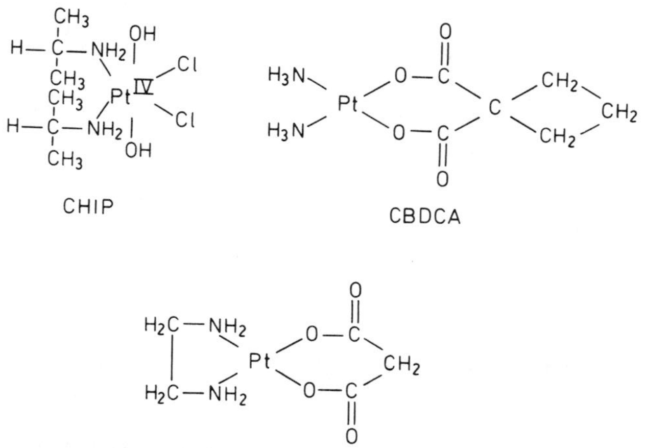


**Fig. 4.** Active analogs of cisplatin being used on a large scale. The two compounds are known best by their abbreviated names. The compound below is (ethylenediamine)malonatoplatinum(II)

To illustrate the type of structure that results in active compounds, the crystal structures of two cisplatin analogs developed in our laboratory are redrawn in Fig. 5. Analysis of the crystal packing has shown that, in the solid state, the hydrogen bonding of the N–H group is clearly intermolecular[31, 32], illustrating at least the possibility that intramolecular hydrogen bonding could indeed occur under physiological conditions.

Knowledge of these basic requirements for an active drug, narrows the area in which one could – successfully – search for new anti-tumor active compounds. Only platinum compounds as described above can be selected on this basis. One should be aware of the fact that platinum compounds that answer the above prerequisites are not automatically applicable as a drug. It does not imply that other platinum (or inorganic) compounds, e.g. with a totally different structure, should not be studied any further for anti-tumor activity. Up till now, too little is known about the cause of the sometimes severe side effects. These effects cannot be predicted on the basis of molecular geometry of the

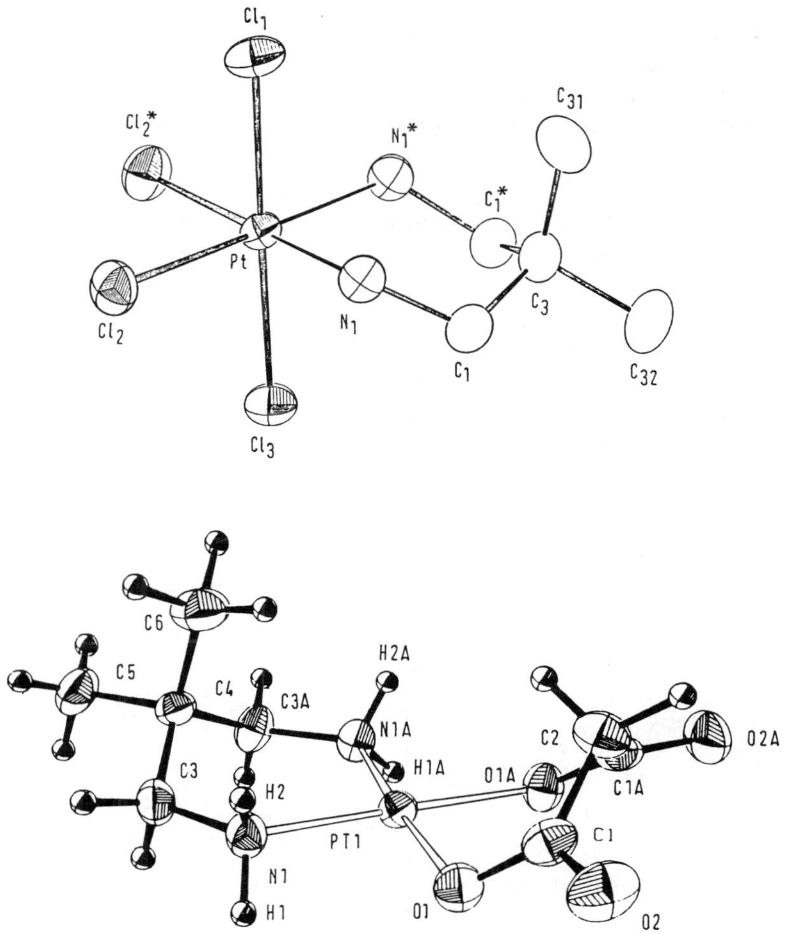


**Fig. 5.** ORTEP-drawings of PtCl$_4$(dmdap) and Pt(malonate)(dmdap), showing the molecular geometry of a Pt(II) and a Pt(IV) compound with antitumor activity

Pt(II) compounds and even the targets (blood, organs, tumor tissue) show a different sensitivity to the drug. It was shown that cisplatin and CHIP differ mainly in kinetics of DNA binding[153].

Finally it should be mentioned that the water solubility of the platinum drugs is an important factor in their activity, as it determines the administration protocol. Research towards new compounds is still ongoing, and recently the structures of several readily soluble derivatives, some of them being Pt(IV) compounds, have been described in detail[32–34].

# 4 Hydrolysis and Kinetics of Platinum Amine Compounds

The reactions that may occur in living tissue and during the transport of drugs through the body are of great importance to understand the mechanism of action. In cases where platinum compounds are used as drugs, an important first reaction appears to be hydrolysis. Most of the present knowledge is based on the hydrolysis of the classical compound cisplatin in aqueous solution. This compound contains coordinated $Cl^-$, which may be hydrolysed depending upon the $Cl^-$ concentration and pH in the surrounding medium. A hydrolysis reaction scheme is depicted in Fig. 6. This scheme is based on the pioneering work of Martin[8, 35] and Lee[36]. At high Pt-concentration and at high pH values, formation of dinuclear and trinuclear hydroxo-bridged Pt-species does occur[16, 17, 20]. However, under biological conditions these species are unlikely to be formed.

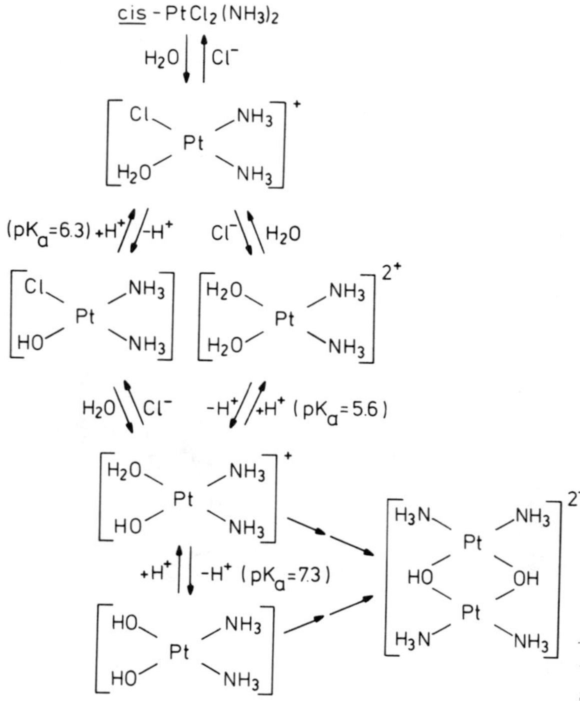


**Fig. 6.** Hydrolysis scheme of cisplatin; the final dimerization (or even trimerization) occurs only at high pH and in concentrated solutions[8, 35]

So, in aqueous solution cisplatin will loose $Cl^-$, and an aqua or/and hydroxo (at high pH) ligand becomes coordinated. In the body, the $Cl^-$ concentration outside cells is rather high (about 100 mM), and therefore hydrolysis is largely prevented there. Inside cells, however, the $Cl^-$ concentration is about 4 mM, promoting the hydrolysis process. According to calculations by Martin[35], the aquated species in the body fluids is present only for a few percent, whereas it amounts to about 50% of the total present inside the cell.

The kinetics of these hydrolysis processes in vivo are not known in detail, but some information is available from in vitro studies. In these investigations cisplatin and analogs were reacted with e.g. nucleotides (such as 5′-GMP). As one would expect, the aqua-species reacts much faster[37] with nucleic acids than the dichloro species. Quantitative kinetic information about these processes is still lacking, however. Generally, dilute solutions with a concentration of the Pt compound of about 0.1–1 mM in the dichloro form, need at least 24 h for completion of the reactions with nucleic acids under ambient conditions (pH 7, room temperature), whereas in case of the diaqua form, the reactions may be completed within a few hours[37–40]. Preliminary investigations with analogs of cisplatin[41, 42] have already shown, that in the case of carboxylate ligands as leaving groups, reactions with nucleic acids proceed much more slowly (despite this fact these compounds are good drugs[17, 40, 41]).

Even less is known about the kinetics of the Pt(IV) drugs, such as CHIP, although evidence is accumulating that binding to DNA only occurs after reduction of Pt(IV) to Pt(II) species[43, 44]. It is evident that much more experimental knowledge in this analytical-physical chemistry area is required. A warning has recently been made[160] to avoid the use of DMSO as a solvent.

## 5 Biological Targets of Platinum Amine Compounds

Evidence is accumulating that DNA is the most important intracellular target of cisplatin. However, DNA is not the only target. Binding to proteins and RNA also occurs, as has been shown by many investigators, see reviews[14, 16, 17]. In addition, it cannot be excluded that damage to cell membranes is also relevant, although no detailed information is as yet available. The possibilities dealing with DNA are schematically depicted in Fig. 7. Evidence for the importance of cisplatin binding to DNA is obtained from the following observations:

- The induction of filamenteous growth of *E. coli* bacteria after treatment with cisplatin[11, 12]. This strongly suggests that cell division (DNA synthesis) is hampered, while cell growth and therefore RNA and protein synthesis continue.
- The prophage induction in lysogenic bacteria under the influence of cisplatin[45].
- The correlation between antitumor activity of Pt-compounds and prophage induction.
- The observation of DNA synthesis inhibition, by measuring the radiolabeled precursor uptake in Ehrlich ascites and human $AV_3$ cells[46, 47], whereas RNA and protein synthesis were hardly affected (see Fig. 8 for a summary).
- The mutagenic properties of cisplatin (resulting in both frame-shift and base-substitution mutations) are also pointing towards a (major) interaction with DNA[48].
- The difference in cytotoxicity of cisplatin for DNA repair proficient and DNA repair deficient bacteria[48] and human cell lines[49, 50].

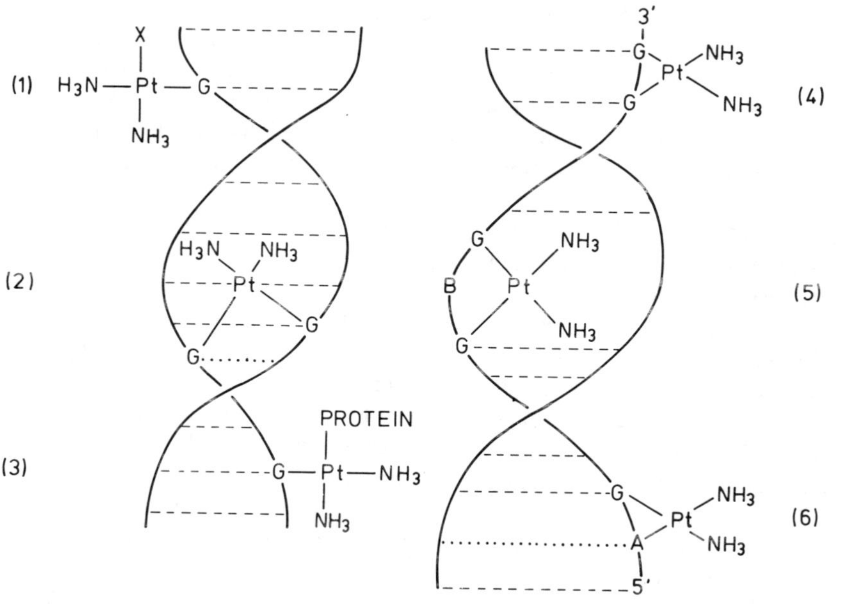


**Fig. 7.** Schematic representation of cisplatin bonding to DNA. *(1)* monofunctional binding ($X = Cl^-$, $OH^-$, $OH_2$); *(2)* interstrand crosslinking; *(3)* protein-DNA crosslinking; *(4)* intrastrand crosslinking between adjacent guanines; *(5)* intrastrand crosslinking between two guanines separated by a third base; *(6)* intrastrand crosslinking at a-AG-unit

However, all this evidence for DNA being the critical target does not necessarily prove that this type of binding is the only reaction leading to cell killing, regression of the tumor and the observed side effects in patients treated with cisplatin. It is quite evident that on the route to the tumor cells, the Pt compounds can interact with e.g. blood proteins and with cell-wall components. These reactions would also contribute to the toxic side effects observed upon treatment of patients with cisplatin and analogs. These reactions have hardly been studied and therefore will not be discussed in the remaining part of the review. From now on we will concentrate on nucleic acid binding. The next

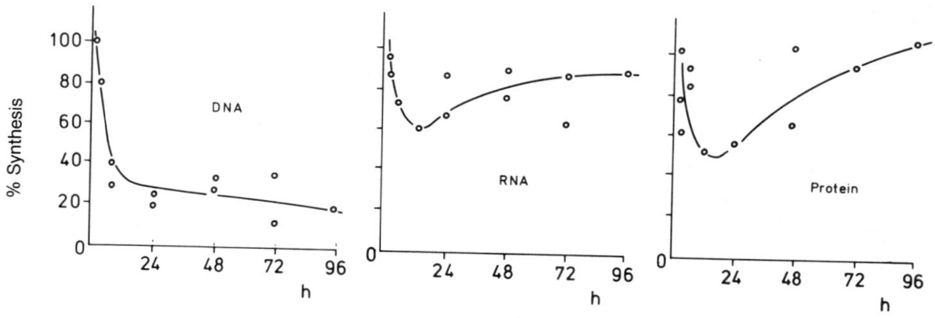


**Fig. 8.** Evidence for DNA being the most likely biological target. Synthesis inhibition as a function of time for DNA (up to 60–70%), RNA (weak temporary effect) and protein (moderate effect, diminishing after some time). After[46, 47]

section will give an introduction about Pt-binding to nucleic acid fragments. The final two sections will deal with in vitro and in vivo reactions with, mainly, DNA.

# 6 Binding of Platinum Compounds to Monomeric Nucleic Acid Fragments

*Nucleobases, Nucleosides and Nucleotides*

The early recognition that binding of cisplatin to DNA is important, has directed many studies towards the area of metal nucleobase interaction. It is impossible to review the many papers that have appeared in this particular field. Only a brief summary will be given below, and for details the reader is referred to previous reviews[18–25, 51–53]. In Fig. 9a, a tetranucleotide with the four common nucleobases of DNA is drawn with the common numbering system. The conformational characteristics and conventions are schematically depicted in Fig. 10. Metal binding can occur in principle at all sites having a lone pair of electrons. However, in practice metal binding occurs only to a relatively small number of sites, i.e. cytosine-N3, guanosine-N7, adenosine-N1 and adenosine-N7. Binding of metal ions to the sugar ring oxygen atoms has rarely been observed in nucleotides and nucleosides. Binding of metal ions to the phosphate groups in nucleotides is quite common, especially for class A metal ions such as Mg(II) and Ca(II), but also binding of Zn(II), Cu(II) and Ni(II) has been observed[53]. In case of Pt(II) – a class B metal –, however, only binding to the nitrogen atoms of the nucleobases has to be

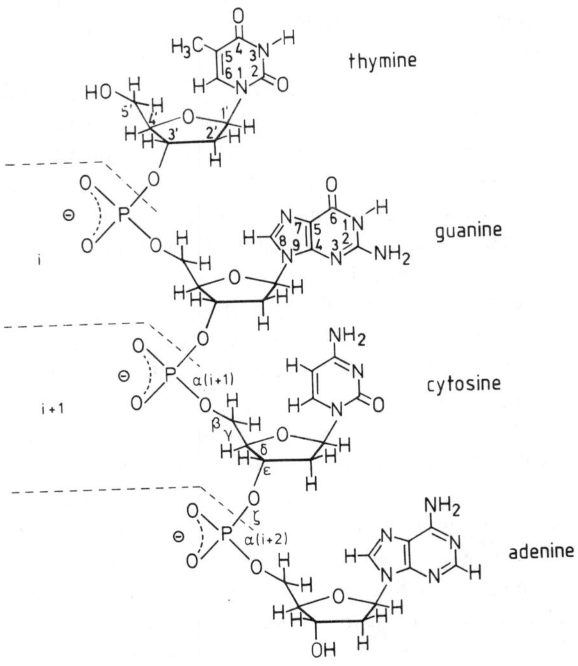


**Fig. 9.** Common nucleobases with used ring numbering system, deoxyribose sugar ring with indications of numbering and conformation

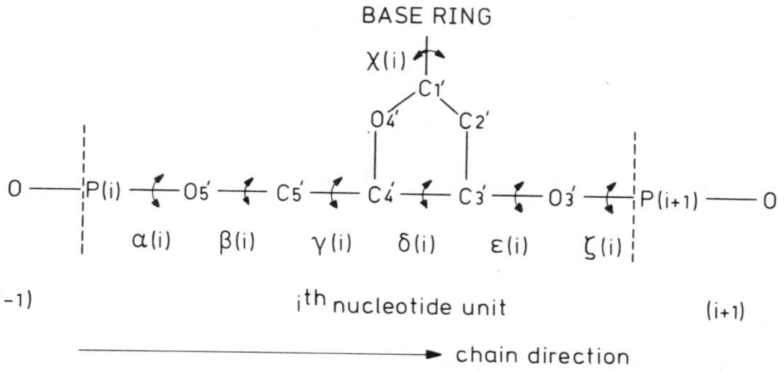


**Fig. 10.** Conformational characteristics and numbering in a oligonucleotide chain

considered, although Pt-phosphate interactions do occur in certain cases[134, 135, 147]. The role of phosphate under physiological conditions seems to be only secondary for coordination, but nevertheless quite important (vide infra for hydrogen bonding).

In Table 1 we have summarised the observed binding sites for Pt(II) to DNA fragments, usually proven by X-ray crystallography in the solid state, and by NMR spectroscopy in solution. It is not possible to review here the great variety of studies dealing with structural and spectroscopic studies on coordination compounds with nucleobases and derivatives.

Especially the various studies that made use of NMR techniques have contributed a lot to a better understanding of the manner in which DNA fragments react with metal ions. High-field NMR spectroscopy is a very useful tool to determine the platinum binding sites in nucleic acid fragments, since crystal structure determinations have been reported only for a limited number of cases. The NMR evidence for platinum binding to nucleobases can be summarized as follows:

- A downfield shift, usually about 0.2–1.0 ppm, of a nearby proton[19, 54].
- The absence of protonation shifts proves platination at this site, i.e. when a metal ion is coordinated at e.g. N7 of guanine, protonation cannot occur anymore at this site, resulting in the absence of a protonation pK at about pH 2. An illustration of such a pH dependent chemical shift pattern of nucleobase protons is given in Fig. 11 a, b. Metal binding can also alter pKa's of sites where no metal is bound, e.g. the decrease of the pKa at N1.
- A characteristic $^{195}Pt$-$^{1}H$ coupling, manifested as satellites around H8 (34% of isotope $^{195}Pt$) is often observed at low magnetic fields. At high fields (i.e. above 1.5 Tesla) this coupling disappears because a scalar chemical shift anisotropy relaxation becomes important[55].

The quite unexpected binding sites, observed in some cases like N3 of GMP (Table 1), are the result of the fact that no alternative binding site was available, or that an excess of Pt[130] was used. In biological systems, all kinds of possibilities are present, and only the kinetically preferred ones are expected to be selected. It should be realised that in case of platinum binding, the kinetic effects dominate the binding, since ligand-exchange reactions of Pt(II) are quite slow[5–7]. In other words: when platinum is attached to a certain site, it may take a long time (hours, days) before the Pt-nucleobase bond is broken. This fact led several investigators almost a decade ago[55–57] to study the competi-

**Table 1.** Summary of observed coordination modes for fragments of nucleic acids and derivatives with respect to Pt(II). For numbering of the nucleobases the reader is referred to Fig. 9

| Base or derivative[a] | Site: | $pK_a$-range[b] | Observed with X-ray or NMR[c] | Comments |
|---|---|---|---|---|
| Guanine | N7 | 2.3 | X, R | Most general coordination mode in many compounds |
| Guanine | N1 | 9.3 | X, R | Observed when N7 is blocked, or with excess of Pt |
| Guanine | N3 | ?[d] | X | Observed with excess Pt |
| Hypoxanthine | N7 | 1.2 | X, R | Similar to guanine derivatives |
| Hypoxanthine | N1 | 8.7 | X, R | Only when N7 is blocked |
| Adenine | N7 | −1.5 | X, R | Most common site |
| Adenine | N1 | 3.7 | X, R | Quite frequently observed |
| Adenine | N6[e] | ?[d] | X | Only very rarely observed |
| Cytidine | N3 | 4.4 | X, R | Many compounds known |
| Cytidine chelate | N4 | ?[d] | X | Only in a N3,N4- with Pt(IV) |
| Uridine, Thymidine | N3 | 9.5 | X, R | Only after deprotonation |

[a] The examples concern derivatives of guanine, comprising guanosine, methyl and ethyl derivatives of guanine, and guanosine monophosphate; Hypoxanthine is a derivative of guanine with the $NH_2$ at C(2) replaced by hydrogen (see Fig. 9); the derivatives comprise 9-methylhypoxanthine, inosine and inosine monophosphate. The adenine derivatives comprise all kind of alkylations at N9 and N7 (methyl, ethyl, butyl), adenosine and adenosine monophosphate. The cytosine derivatives comprise 1-methyl and 1-ethyl derivatives, cytosine and cytosine monophosphate. The derivatives of uridine and thymidine concern the 1-alkyl derivatives, the 1-ribo and 1-deoxyribose derivatives, as well as the monophosphates

[b] The values of the $pK_a$ were taken from compilations of Martin[25)]. Values above 7.0 indicate that under normal conditions these sites have a hydrogen atom and will not coordinate without loosing the hydrogen

[c] X = crystal structures known; R = structures proven by NMR studies

[d] $pK_a$ values not accurately known

[e] N6 is the exocyclic nitrogen, after deprotonation

tion kinetics, in which a mixture of nucleobases, nucleosides and/or nucleotides was used. From these studies, it became apparent that the guanine-N7 binding is strongly preferred under a variety of conditions, followed by Adenine-N7, Adenine-N1 and Cytidine-N3. In addition, the presence of a 5′-phosphate group in a mononucleotide results in a faster reaction than in the presence of a 3′-phosphate group[40)]. Recent theoretical studies[133)] have confirmed the strong thermodynamic preference for guanine-N7. The kinetic preference might be related to the attractive hydrogen bonding at O6 (see below).

It should be realised that variation of pH may change the kinetic preferences, especially at high pH (above pH 10), when the guanine-N1 may loose its hydrogen and become accessible for platination. Also thymidine may bind to platinum at high pH. Removal of platinum from the nucleobase after binding is a slow process (vide supra). However, with certain strong ligands of the class B type, such as $CN^-$ or thiourea, the Pt–N bonds may be broken more easily[17)]. For example with an excess of cyanide, the platinum may be removed as $Pt(CN)_4^{2-}$.

Migration of platinum from one site of a nucleobase to another is quite uncommon, and not likely to occur under ambient conditions. Only in a few cases has isomerisation of N7 to N1 and N1 to N7 in Pt-nucleobases been observed[58, 59)]. In these cases the ligands

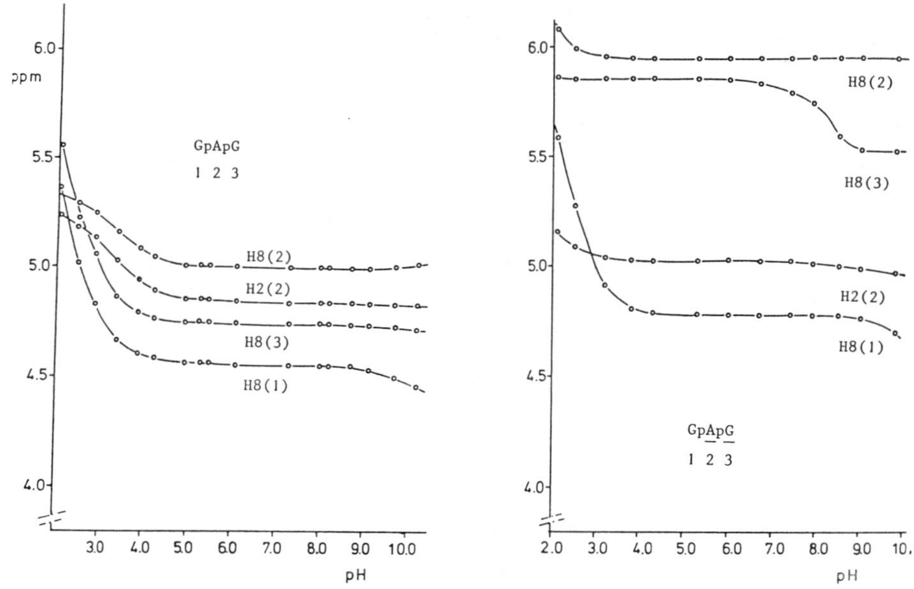


**Fig. 11a, b.** Evidence for N7 binding of cisplatin, through monitoring the pH-dependence of the chemical shifts of the purine protons. **a)** The free trinucleotide GAG with its H8 signals (for each base) and H2 signal (for Adenine). Effects of N7 protonation, upon lowering the pH, are clearly seen for H8(1) and H8(3) at pH = 3, and somewhat less pronounced at H8(2) and H2(2). **b)** The trinucleotide GAG with a *cis*-Pt unit chelated to the N7 atoms of A(2) and G(1) or G(3). The observations are:

1. The H8 resonances of A(2) and G(3) are shifted downfield upon coordination.
2. The signal of H8(3) shows a pH effect at pH 8, which is due to protonation at N1(3).
3. Protonation effects for N7 at pH 3 have disappeared, except for H8(1), because this base has no platinum at N7 (and can still be protonated)

on platinum (N–H donor) and the oxygen-6 of the nucleobase (H-bond acceptor) probably facilitate such an intramolecular process.

The role of the oxygen atom on guanine, O6, has been the subject of much debate in the past. It had originally been assumed that a chelate with *cis*-$Pt(NH_3)_2^{2+}$ and the N7–O6 groups of guanine (see Fig. 9) was an important end product[60], possibly explaining the difference between the cis- and trans-isomers. However, this binding is now generally accepted to be highly improbable and irrelevant[17], though a hydrogen-bond accepting role of the O6 group is quite likely[31, 38], and in fact clear from several X-ray studies[140] of solid products.

Cisplatin and analogous compounds have two reactive binding sites, and therefore usually can react with two nucleobases. In many cases the 1:1 product has not even been observed[38]. Observation of 1:1 products, however, is possible by taking sterically-hindered asymmetric diamines[40], by lowering the temperature, by using an excess of cisplatin, or by adding an excess of a competing ligand[38] such as $Cl^-$. Using $N^2$,$N^2$-dimethyl-9-methylguanine(Dmmg), it was possible[141] to crystallize the 1:1 product *cis*-$[PtCl(NH_3)_2(Dmmg)]PF_6$. For the study of a 1:1 end product, often [PtCl(dien)]Cl (see Fig. 1) is used.

An important observation, reported by Cramer[61] and by Marcelis[40], was that in certain cases the rotation about the Pt–N7 bond is slow on the NMR time scale, as a result of steric hindrance in the coordination sphere of platinum. From the fact that DNA reacts with the two available coordination sites of cisplatin, the species formed upon the first binding step should be able to undergo fast rotation about the Pt–N bond in a "search" for a second binding position on the DNA[19]. The fact that with small amine ligands on platinum, i.e. the anti-tumor active compounds, such a rotation is always fast, confirms the earlier observation (Sect. 3) that amine ligands with at least one N–H group yield active compounds. Inactive compounds with bulky amine groups not only have different dynamics (rotation about the Pt–N bond), but also result in different structures of the formed adducts[40, 61].

*Differences between Cis and Trans Isomers*

It has been known for some time that *trans*-$PtCl_2(NH_3)_2$ binds also to DNA[14, 18]. It is likely that the first step does not differ very much from that of cisplatin. With mononucleotides, *trans*-$PtCl_2(NH_3)_2$ also easily forms 1:2 adducts, and these differ hardly from those derived from cisplatin[38]. Also the kinetic differences of the cis- and the trans-isomer hardly differ ($Cl^-$ and $NH_3$ are close in the trans-labilizing series[5–7]; vide supra). The biological effect of *trans*-$PtCl_2(NH_3)_2$ is quite interesting and DNA binding studies deserve detailed study[62, 151]. Since it is clear that the binding affinities to the nucleobases are quite comparable, the different anti-tumor activity must originate from differences related to the bifunctional binding. Comparative studies with trans- and cis-platin on single-stranded and double-stranded oligonucleotides will certainly contribute to the explanation of these differences (vide infra).

*Role of Hydrogen Bonding*

The influence of the N–H group in the amine ligand of the Pt-compound (see Sec. 3) may have an indirect role in the binding to nucleic acid fragments and to DNA. On the one hand it could have a directing effect on the approach to the DNA, by having a H-bond interaction with guanine-O6. This possibly explains the kinetic preference of platinum compounds for guanine-N7. On the other hand, after formation of a bis-nucleotide adduct, it may have a H-bond interaction with a phosphate group, as has recently been found[63] in a crystal structure determination of *cis*-$Pt(NH_3)_2$(pGG-*N7,N7*) and was expected from molecular mechanics calculation[64]. Several crystal structures of mononucleotides have shown earlier the great importance of hydrogen bonds in the solid state[23–25, 51, 52]. The role of hydrogen bonding will be discussed in more detail in the next section.

## 7 Binding of Platinum Amine Compounds to Oligonucleotides

*General Aspects*

The knowledge that cisplatin has two binding sites available and that binding to nucleobases in DNA may occur, has led to investigations on Pt-binding to dinucleotides. In these compounds the second step of the binding (chelation) becomes possible within the same molecule. Of all 16 possible combinations of nucleobases (GG, AG, GA, CG, GC, TG, GT, AA, AC, CA, AT, TA, CC, CT, TC, TT) and other variations (i.e. ribose or deoxyribose; with or without a terminal 3'- or 5'-phosphate group) only a few have been studied in great detail. In addition to dinucleotides, larger oligonucleotides of defined sequences (up to dodecamers) have been used to study the binding of cisplatin, the trans-isomer and many other compounds.

Questions addressed in these studes are:

- Is there a kinetic preference for any of these combinations (i.e. will a certain sequence be recognized)?
- What are the binding sites at the nucleobases?
- Are products with different sugar conformations and/or base orientations (isomers) formed?
- Can a detailed structure be determined, either in the solid state or in solution?
- What is the effect of the sugar ring on binding?
- What is the effect of phosphate groups on the kinetics and structure of the formed chelate?

Although work in this area is still ongoing, by others and in our laboratories, some important results have already come out[17–19, 39, 63–96] and will be discussed here.

*Dinucleotides*

Starting with a discussion of the kinetic preferences, it should be pointed out first that a dinucleotide (which is single stranded and largely unstacked) most likely will react differently compared with the same nucleotide when part of a double-helical, stacked DNA sequence. Therefore, the outcome of a comparative study on dinucleotides cannot be extrapolated directly to larger single-stranded, or double-stranded DNA structures. Preliminary results obtained by van der Veer[59] in competition experiments with d(XpG) dinucleotides, suggest a certain directing effect of the 5'-neighbouring base (C, T or A). The reaction rate of the dinucleotides with $[Pt(dien)CL]^+$ decreases in the sequence C, T, A.

A second point of interest is the question of the stoichiometry. For example when cisplatin is reacted with the dinucleotide CG in a 1 : 1 ratio, the resulting product may be a 1 : 1 adduct, but also *cis*-$Pt(NH_3)_2(CG\text{-}N7)_2$ may be formed (leaving unreacted cisplatin in solution). A detailed study of this system by Chottard[86] has made clear that in dilute solutions the 1 : 1 product is formed; only when the concentration of CG is high, is the 1 : 2 product formed. This agrees with earlier observations of Jordanov[66] and Kidani[67] for the reactions of cisplatin and CG and GC, respectively.

The binding sites of the nucleosides in all these studies are not unexpected, i.e. guanosine is invariably coordinated at N7, cytosine always at N3, whereas in case of adenosine both coordination at N1 and at N7 has been observed, depending upon the conditions[42, 68, 70, 75].

A very interesting observation, first made by Chottard et al.[70], deals with the occurrence of conformational isomers after chelation of cisplatin to a number of dinucleotides. In certain cases these isomers are rapidly interconverted, but in many cases they are easily separated by chromatographic techniques, and allow characterisation by NMR techniques[70, 72, 75, 92]. For such an analysis, application of high-field NMR in conforma-

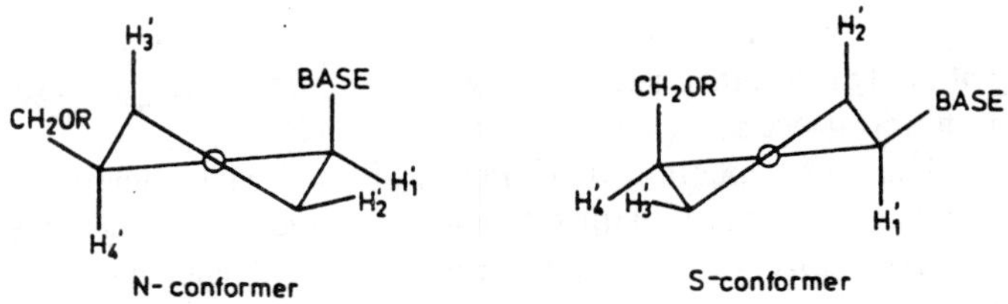


**Fig. 12.** Projection of the two extreme conformers of the furanose ring in nucleic acids. *S* occurs in B-DNA, *N* occurs in RNA

tional analysis is a prerequisite, and for a review about this subject the reader is referred to a detailed treatment by Altona[93]. Only the most important characteristics are summarised here. First one should know that the sugar rings are engaged in a so-called N–S equilibrium through pseudo-rotation[93], and the two extreme conformers are depicted in Fig. 12. In B-DNA an S-type sugar conformation is present, whereas in A-DNA and in RNA the N-type conformation occurs[93]. In Z-DNA the deoxyribose rings have, alternately, S and N conformations. Secondly, for detailed conformational studies, one needs parameters to describe the helical strands in larger fragments. However, such a detailed description of the range of values for the angles in Fig. 10 is beyond the scope of the present review. Finally, the value of the angle $\chi$ in Fig. 10 is dealing with the conformation of the base with respect to the sugar ring (see Fig. 13). For values of $\chi$ between 20 and 100° term *syn* is used, whereas for values between –60 and +180° the term *anti* is used. In most cases NOE measurements clearly indicate which of these orientations is present[72, 78, 83, 86, 93].

Apart from NMR, other techniques – such as CD[68, 75] and IR[142] – may be used, often complementary, to study the structure and conformation of small oligonucleotides.

It should be mentioned here that when two nucleobases are cis coordinated to platinum, two orientations of the bases can occur, i.e. head-to-head and head-to-tail (see

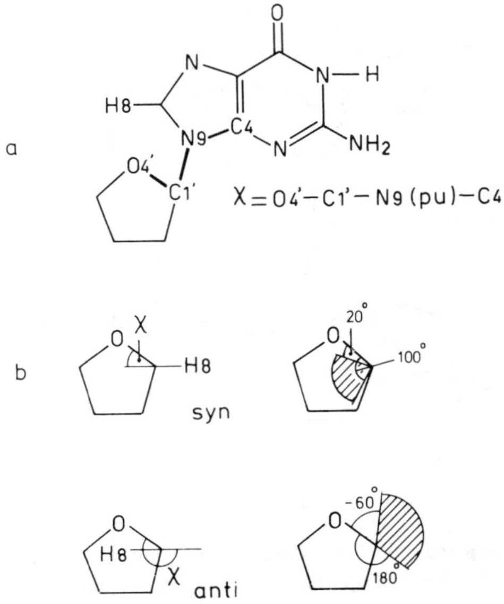


**Fig. 13.** Schematic representations of the *syn* and *anti* conformational ranges in nucleosides, as expressed by the torsion angle $\chi$

Fig. 14). With mononucleotides in most cases head-to-tail orientations have been found with only rare exceptions[131, 158]; with oligonucleotides, however, a head-to-head orientation – as it occurs in DNA – is more likely.

The above-mentioned *syn-anti* conformational isomerism is present in the reaction products of cisplatin with CG and pCG and will be discussed briefly[86]. The binding process clearly is a two-step reaction with initial binding to the N7 of G. After the second reaction step a mixture of C*(anti)*-G*(anti)* and C*(syn)*-G*(anti)* is observed in a ratio 20 : 80 and with C–N3 forming the second coordination site at platinum. The isomers have been separated by HPLC[86] and were found to re-equilibrate slowly at room temperature. From the values of the activation parameters[86] and from comparative studies with the bulkier 5-methylcytosine derivative[94], it has been concluded that the equilibration process is in fact a conformational isomerisation through the rotation of the cytosine about its glycoside and Pt–N3 bonds. Similar observation were found for GA and AG[149].

It is interesting to note that in this case no significant differences are observed for ribose and deoxyribose sugars. Also the differences between cisplatin and the more reactive hydrolysed product (see Fig. 6) are small. The effect of a 5′-phosphate group (i.e. the difference between CG and pCG) on the reactivity appears to be negligible. However, for the effect of an extra 5′-phosphate group next to a guanosine, see below.

Using high-resolution NMR techniques it appeared possible[72] to determine the most likely structure in solution for the adduct *cis*-$Pt(NH_3)_2(d(GG))^+$. Complete assignments of the proton resonances, decoupling experiments and computer simulations yielded a structural model, stereoscopically depicted in Fig. 15, with the following characteristics:

- The two bases, coordinated through N7, are oriented head-to-head with a dihedral angle for the bases of about 60° (cf. Fig. 14)
- The sugar ring at the 5′ side of d(GG) has changed to the N-conformer (compared to the S conformer in free d(GG); cf. Fig. 12).

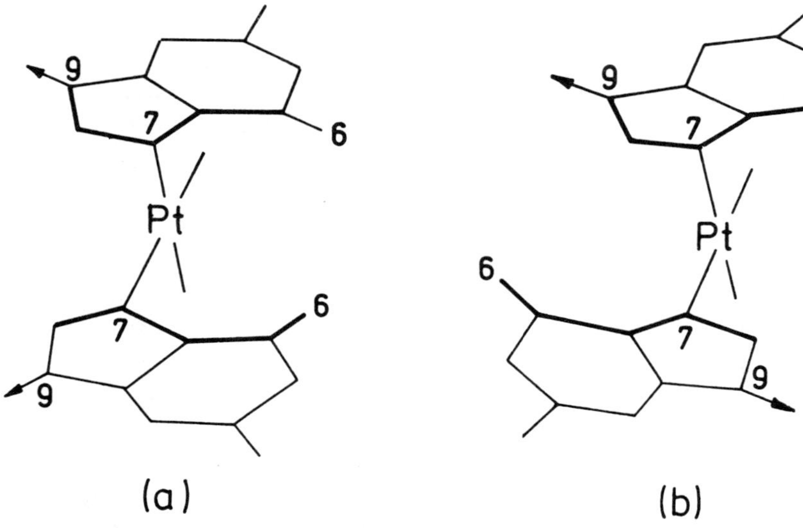


**Fig. 14a, b.** Schematic representations of head-to-head **(a)** and head-to-tail **(b)** orientation of two guanine bases coordinated to a *cis*-Pt unit

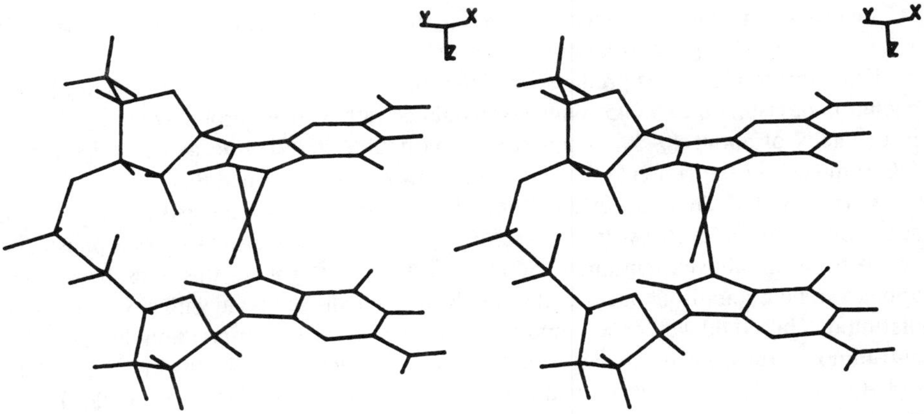


**Fig. 15.** Stereo drawing of the most likely conformer of the cation *cis*-$Pt(NH_3)_2$(d(GG-*N7,N7*))$^+$ in solution as determined by high-resolution NMR techniques

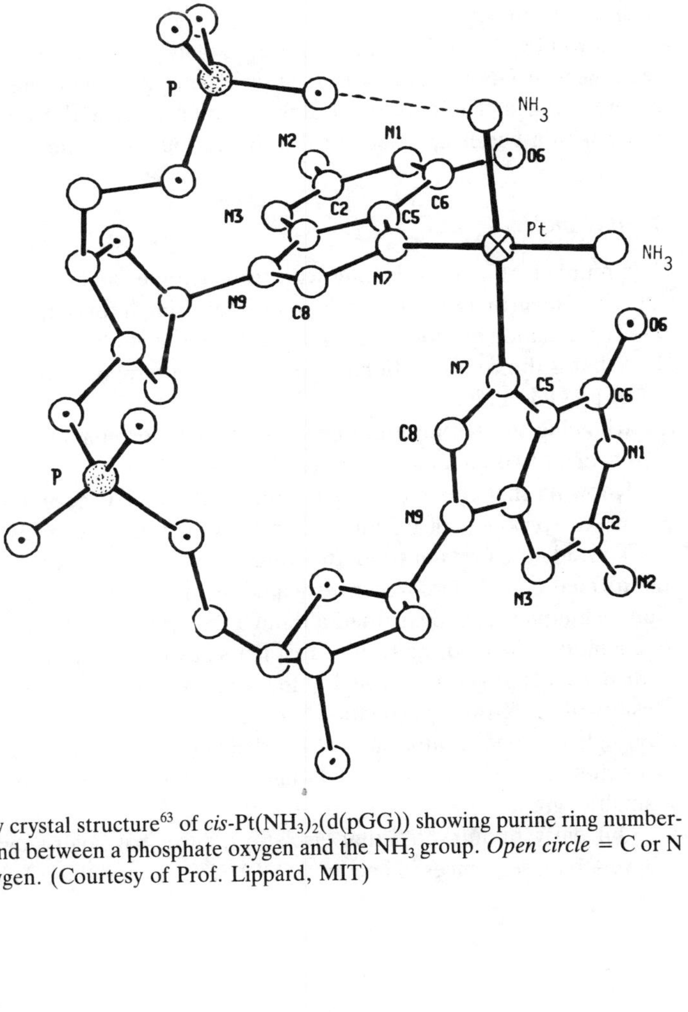


**Fig. 16.** Part of the X-ray crystal structure[63] of *cis*-$Pt(NH_3)_2$(d(pGG)) showing purine ring numbering, and the hydrogen bond between a phosphate oxygen and the $NH_3$ group. *Open circle* = C or N atom; *dotted circle* = oxygen. (Courtesy of Prof. Lippard, MIT)

– The other conformational characteristics of d(GG) are hardly changed upon chelation to cisplatin. The $\chi$-values amount to 115° (i.e. anti configuration).

From this model it was predicted[72, 94] that the distortion of the DNA structure, after chelation of cisplatin to a GG sequence, would be rather small, probably yielding a kink in the helix of about 40–70°. Later studies on double-stranded decanucleotides with a GG-sequence in the center[80, 91, 148, 150] (vide infra) have confirmed this.

Very recently, Sherman et al.[63] have been successful in solving the X-ray crystal structure of the adduct *cis*-$Pt(NH_3)_2$(d(pGG)). This species differs by one phosphate group from the above-mentioned adducts of d(GG). The role of this phosphate group appears to be a stabilising one, i.e. it is involved in a hydrogen bond with a $NH_3$ ligand at platinum. This could well be an important explanation for the observation that platinum anti-tumor drugs need an acid N–H group to donate a hydrogen bond. The $O_3P{-}O\cdots H{-}NH_2$ interaction could stabilize and induce the distortion of the DNA, thereby hampering or blocking the replication processes (vide infra). The most important finding of this study, however, appears to be that the solid-state geometry is substantially the same as the one in solution[72]. Part of the solid state structure is redrawn in Fig. 16, based on the coordinates of Sherman et al.[63]. The analogy with Fig. 15 is striking indeed. Finally it should be mentioned here that such a distortion for a GG-chelate has also been calculated using molecular mechanics[64], and that Kidani[154] found a similar structure with other amines.

As will be shown in a later section, cisplatin shows a very strong preference for GG-sequences in DNA, i.e. a G in a GG sequence appears to have a preference above the other guanines. The reasons for this are still subject of debate, but may have something to do with a better approach of the *cis*-Pt group in an electrostatic manner.

*Trinucleotides*

The number of studies dealing with trinucleotides and cisplatin or derivatives is rather limited. Nevertheless, they yield very interesting information.

Three major questions are adressed in these studies:

1. What is the conformational effect of a third terminal base on the chelating properties of a –GG– unit?
2. Knowing the strong preference of cisplatin for guanine-N7 binding, does chelation occur to two guanines separated by a third base?
3. Knowing that chelation to AG-units, but *not* GA, occurs in vitro (see below), what are the products of reaction when GAG is treated with cisplatin?

To study the first question, the trinucleotide d(CGG) was treated with cisplatin[89] and the formed GG-chelate was investigated by high-resolution NMR with the aid of simulation techniques. It appears that, although the central G has a N-type sugar ring as a result of chelation, the cytosine residue is still able to stack on top of G(2). The chelate formed with the G(2)G(3)-unit is roughly the same as the one found with d(GG) (see above). Results of an X-ray investigation[96] confirm that C(1) is indeed stacked on top of G(2) although a unique, complicated H-bonded network is present. This in fact indicates that – after chelation of cisplatin to a GG-unit – the helical stacked structure is distorted only to a small degree.

Our early findings[48, 97] that cisplatin induces base-pair substitutions in –GCG– and –GAG– base sequences in bacterial DNA *(E. coli)* and thus plays a role in the mutagenic properties, have prompted us to study cisplatin interactions with d(GCG)[76], d(GTG)[98], d(GAG)[99] and d(pGGG)[59]. More recently we have also used analogs of cisplatin[85, 98] and some inactive compounds, such as $PtCl_2$(tmen) and *trans*-$PtCl_2(NH_3)_2$. It appears that the *cis*-$Pt(NH_3)_2$ unit may form a chelate with the –GBG– unit through G(1) and G(3). With other *cis*-Pt compounds very similar structures have been found, although a slightly different conformation was found with the inactive $PtCl_2$(tmen), probably due to steric reasons[85]. Very surprisingly, the inactive trans isomer of cisplatin also forms a chelate with d(GTG), as clearly shown[98] from NMR analysis and enzymatic digestion techniques (see below). A major structural difference with the cisplatin chelate appears to be a conformational change the 5′-end guanine sugar ring.

Recent investigations on d(GAG)[99] have shown that cisplatin also has a tendency to form an AG-chelate (20%, compared with 80% of a chelate through both guanines). Careful analysis[99], using also Pt(dien)$Cl^+$ as a reference, have made clear that the first step is an attack at G(3) (65%), rather than at G(1) (35%). Once binding to G(3) has occurred, the second step can be either at A(2), or at G(1). Similar studies on GGG were hampered by solubility problems[59], but pGGG reaction with cisplatin results in a major product being[59] *cis*-$Pt(NH_3)_2$d(pGGG)-*N*7(1), *N*7(2).

### *Larger Single Stranded Oligonucleotides*

The first studies with larger oligonucleotides in fact have dealt with self-complementary (i.e. forming a duplex with itself) tetranucleotides[69] and hexanucleotides[71, 73], viz. d(CCGG), d(AGGCCT) and d(TGGCCA). In all these studies it became immediately clear that the double helix is disrupted upon chelation of the *cis*-Pt unit at a GG site. Again NMR proved to be a very useful technique to study these reactions and their products. Another technique that can be applied to determine the binding sites of *cis*-$Pt(NH_3)_2^{2+}$ in oligonucleotides, is based upon the application of enzymes that digest the DNA (or oligonucleotides) to smaller fragments, some of which contain platinum species. This technique was first introduced[100] for the analysis of cisplatin-treated salmon sperm DNA. Since then it has also been applied for the analysis of platinated oligonucleotides[79, 82, 84, 98, 99].

More recent studies on ss oligonucleotides are dealing with[78, 82–84, 142] d(ATGG), d(CCATGG), d(GGCC), d(GCGC), d(pGCGC), d(TGGT) and d(pCGCG). Because larger oligonucleotides are needed to maintain a double-stranded structure after platination, more attention has been given recently to octa, deca, undeca and even larger oligonucleotides[80, 87, 88, 90, 91]. Since these larger ones are usually studied in connection with a (self)-complementary strand, their results will be treated in the next section.

### *Double-Stranded Oligonucleotides*

As discussed above, the first investigations on oligonucleotides made clear that – at least up to hexamers – platination results in disruption of the double-helical structure. One should realise, however, that in these self-complementary oligonucleotides two Pt groups will chelate. Therefore, we have undertaken a study with a duplex fragment, having a centrally located GG-unit in only one strand[80, 87, 91, 94]. The performed reactions are depicted schematically in Fig. 17. Upon reaction of cisplatin with the strand d(TCTCGGTCTC) a high yield of the GG chelate is formed, as expected; subsequent

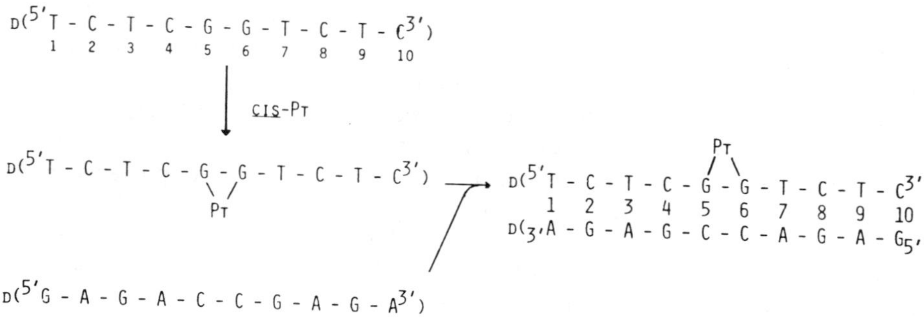


**Fig. 17.** Reaction scheme in the study of the duplex decanucleotide *cis* Pt lesion

addition of the complementary strand d(GAGACCGAGA) apparently results in duplex formation[80]. Detailed analysis of proton and phosphorus NMR spectra[91] and consideration of the CD spectra (see Fig. 18), has resulted in the following conclusions:

1. The double helix is somewhat destabilized, compared with the unplatinated duplex, as seen from the decrease in melting temperature of about 10–20°.
2. Observation of the imino-proton resonances – albeit somewhat shifted and broadening at low temperatures – of the central G:C base pairs, indicates that hydrogen bonding (Watson-Crick type or related) is still possible after platination.
3. Careful analysis of the chemical shifts and coupling constants after platination, suggests a distortion of the double helix which is rather small and which can – at the moment – best be described as a kink in the helical axis of about 40–70°, without a large change in overall helical unwinding.
4. Comparison of CD spectra and $^{31}P$ NMR spectra of platinated DNA (from several sources) and the platinated ds decanucleotide, strongly suggests similar distortions in both cases[87, 94, 101].

In an independent study, using the platinated octanucleotide d(GATCCGGC) and the complementary decanucleotide strand d(GCCGGATCGG), Chottard[88] arrived at similar conclusions. Studies with larger GG-containing fragments are in progress. A detailed study on d(GCCĜĜATCGC):d(GCGATCCGGG) has shown that the GC base pairs attached to platinum melt first[150].

In a related piece of work we have platinated the undecamer d(TCTCGTGTCTC) and have concluded that a chelate using both guanosines (a GBG chelate) is again formed, despite the T in between them[90]. Now addition of the complementary d(GAGACACGAGA) again results in duplex formation. However, the CD spectrum has significantly changed compared with the unplatinated duplex (cf. Fig. 18b) and this change is different from the change in the case of the decanucleotide (Fig. 18a). Detailed study of the proton NMR spectrum in the imino region indicates that two H-bridges between the bases have disappeared and that the helix melting temperature has decreased by about 25°. Apparently the central T:A base pair cannot be formed at all, since T is bulged out as a result of chelation of the guanines to *cis*-Pt[90]. Addition of the almost complementary decanucleotide d(GAGACCGAGA) did not result in a better duplex formation[94].

Recently also Kidani[95] has shown – by using enzymatic digestion and HPLC – that cisplatin binding to self-complementary d(GGTCGACC) and to d(CGGATCCG) results mainly in *cis*-Pt-GG adducts.

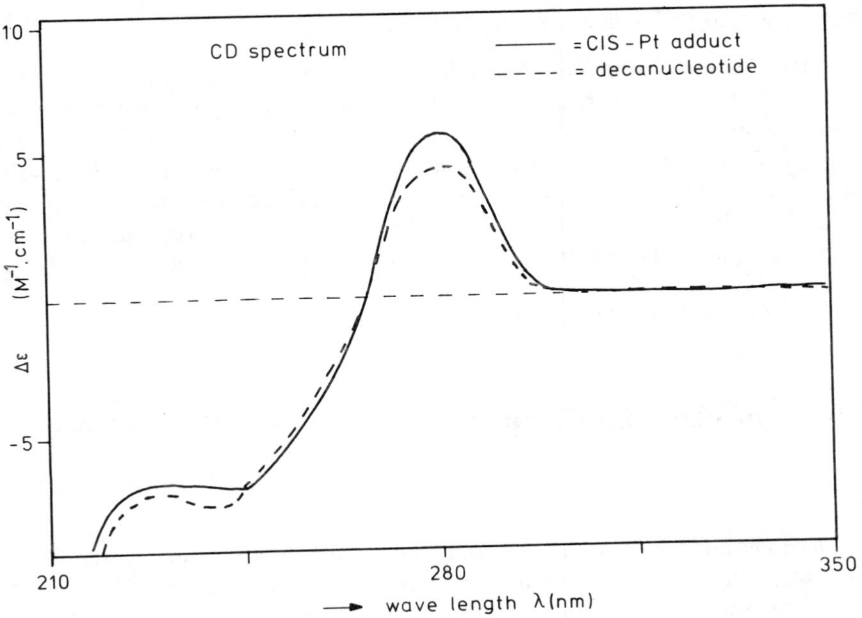


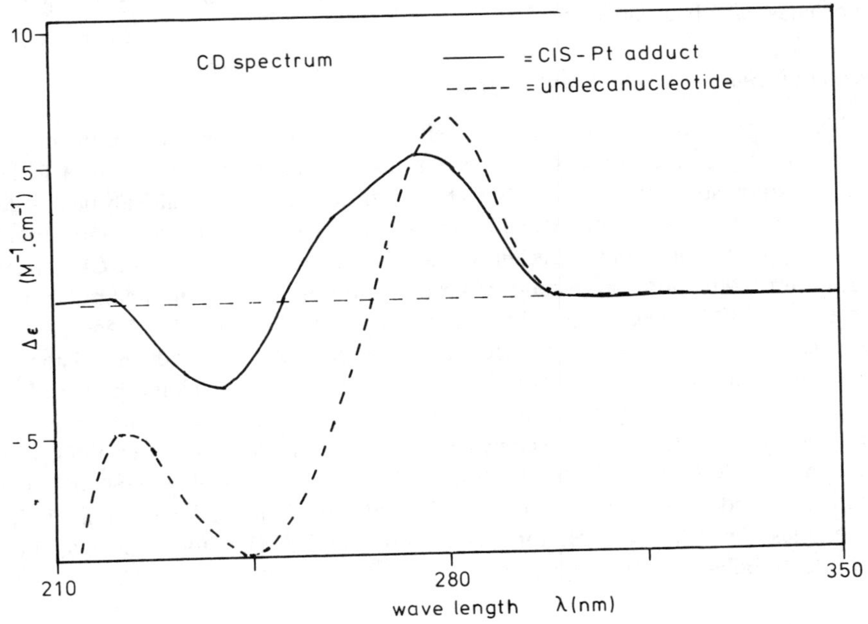


**Fig. 18.** CD spectra of double-stranded decanucleotides and undecanucleotides with and without *cis*-Pt chelation at the central guanines of the strands d(TCTCGGTCTC) and d(TCTCGTGTCTC) respectively

*Conclusions About Pt-Binding to Oligonucleotides*

From all the above-mentioned results the following general conclusions can be drawn:
1. When adjacent guanines are present in an oligonucleotide, cisplatin chelates to these at the N7 positions.
2. When GBG sequences are present in an oligonucleotide, cisplatin chelates also to both guanines; when the central base is adenine then also an AG chelate is formed.
3. Double-stranded oligonucleotides may be formed with the above-mentioned platinated GG- and GBG-containing oligonucleotides (at least 8 units seem to be required).
4. The distortion of the double helix is rather small in case of GG-chelation by cisplatin.

# 8 Binding of Platinum Compounds to DNA and RNA in Vitro

*Scope*

As soon as it became apparent that the cisplatin-DNA interaction is an important event in the cytostatic process, investigations were started using DNA from several sources. In this review attention will be given to recent spectroscopic studies dealing with DNA and RNA binding, with special emphasis on the molecular details, as well as on quantitation of the several adducts formed. Other aspects have been treated in numerous reviews[10, 12, 17, 22], to which the reader is referred. Most studies have dealt with cisplatin, although recently the differences between *cis*- and *trans*-$PtCl_2(NH_3)_2$ in binding to SV40 chromosomes have been presented[62].

*Spectroscopic Studies on DNA and Polynucleotides*

Upon binding of cisplatin to DNA in vitro (either ss or ds) one may monitor the differences in spectroscopic properties. A very sensitive technique has been shown to be circular dichroism spectroscopy[70, 103]. In fact, early observations[103, 104] already dealt with the enhancement of the ellipticity at 275 nm, indicating the occurrence of intrastrand cross linking. Our results on the decanucleotide (see above) show the same CD effects (Fig. 18) after platination, suggesting similar distortions in the decamer and in DNA.

More recently it has become possible to use $^{31}P$ NMR[87, 101] to study DNA and – under comparable conditions – double-stranded oligonucleotides. As already discussed above, the first results indicate that similar distortions occur in DNA and oligonucleotides. It is even possible now[105], applying special proton NMR techniques, to use imino proton signals to study the distortions induced by cisplatin, as nicely shown for the ribopolymer poly(I) : poly(C) by Marzilli[105]. In all these studies one should realize, however, that under in vivo conditions the loading of the DNA by cisplatin is much lower than in the DNA samples, which are suitable for spectroscopic studies. Therefore, one should be careful in extrapolating such results.

*Binding to RNA*

Easier than is the case with DNA, the smaller RNA molecules may be purified and even crystallized as a molecule. This allows the study of cisplatin binding to such molecules in the crystalline state. Two studies have been reported recently[106, 107] both dealing with yeast transfer $RNA^{Phe}$. Although RNA does not have a double helical structure like DNA, many regions with base pairing do occur, resembling somewhat the structure of DNA. Careful comparison of diffraction data from pure RNA and platinated RNA's should give positions for Pt binding.

Rubin et al.[106], by soaking their crystals with cisplatin, found only monofunctional binding at certain guanine-N7 sites. After soaking with the trans isomer, the same binding was found, but also binding at adenine-N7.

Dewan[107], on the other hand, soaked the crystals with hydrolyzed cisplatin – the most likely species inside cells – and found that bifunctional binding can occur. He concluded that chelation at GG and – to a lesser extent – at AG (but not at GA) are the predominant binding events.

*Binding to DNA*

A different approach to study the cisplatin binding positions in DNA is based upon degradation of high-molecular weight platinated DNA, and analyzing the fragments – after separation – by a special technique, such as NMR. In our laboratories a method has been developed[100, 108] to degrade cisplatin-treated DNA with enzymes to mononucleotides and Pt-containing mono- and dinucleotides. This mixture was subsequently separated on the basis of their net charge on an anion-exchange column. By applying this method, four reaction products of cisplatin with salmon sperm DNA were isolated, together comprising at least 90% of the total platination[108]. The structures of these adducts have been unambiguously proven by proton NMR spectroscopy and Pt-analyses with atomic absorption spectroscopy (AAS). In all cases the Pt was linked to the N7 atom of guanine or adenine and no indications have been found for binding at adenine-N1 or cytosine-N3 at the level of platination studied (up to 6 Pt atoms per 100 nucleotides; $r_b$ = 0.06). One of the adducts, i.e. $Pt(NH_3)_3(dGMP)$, appears to be derived from cisplatin that is monofunctionally bound to guanine (see Fig. 7). The fourth ligand in this digestion product is $NH_3$, which originates from the ammonium hydrogencarbonate buffer used in the process of inactivation after platination[109]. The other adducts are all derived from bifunctional cisplatin binding to DNA.

The formation of intrastrand crosslinks on GG and AG base sequences in DNA could be deduced from the presence of the adducts *cis*-$Pt(NH_3)_2(d(pGG))$ and *cis*-$Pt(NH_3)_2(d(pAG))$ in digests of platinated DNA. However, no indications for the presence of GA adducts could be found, and therefore crosslinks with a GA sequence are unlikely.

The formation of interstrand crosslinks in DNA, also demonstrated with alkaline elution techniques[14, 50, 110], could be related to the presence of the adduct *cis*-$Pt(NH_3)_2(dGMP)_2$. This product might originate from two guanines on different strands. The same Pt-digestion product, however, is also obtained from intrastrand crosslinks on two guanines separated by one or more nucleobases (see Fig. 7). The occurrence of this type of crosslinks in DNA was deduced from mutation studies in bacteria[48, 97], and proven in oligonucleotides (see above).

In Fig. 19 the induction of the various cisplatin adducts in salmon sperm DNA at 37° is given. The data, as measured with AAS, show that from the onset of the reaction the spectrum of adducts remains unchanged and that cisplatin has a strong preference for binding at the sequence pGG. It was calculated[108] that this adduct comprises 60–65% of

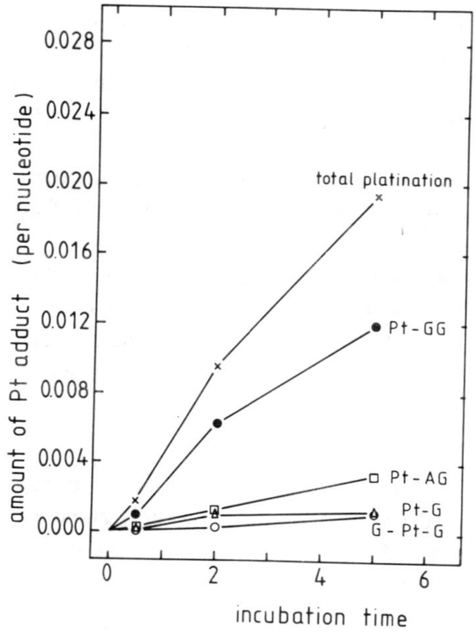


**Fig. 19.** Amounts of the platinum adducts, as a function of incubation time (h), found upon treatment of salmon sperm DNA with cisplatin at 37°

the total platination, whereas on the basis of a random distribution of guanines in this type of DNA a value of only 37% would be expected if all guanines would be equireactive. From the almost constant and small amounts of monofunctionally bound cisplatin (see Fig. 19) it may be concluded that the reaction of cisplatin with DNA is mainly proceeding via an, almost simultaneous, chelation of the platinum with two N7 atoms of purines, preferentially located in a GG-sequence.

Studies of cisplatin bound to DNA have also been reported by others. Eastman[111, 112] has used radiolabeled $PtCl_2(en)$ to detect and quantify the adducts after enzymatic hydrolysis. Johnson[113, 114] has used an acid-catalyzed depurination followed by electrophoresis. Rahn[115] used radiolabeled platinum compounds for detection, whereas Olinsky and Walter[116] have used the $SnCl_2$ method to quantify the amounts of the various cisplatin-DNA adducts.

In all cases products containing guanine, and guanine together with adenine were found. Assuming that the first step will always be a binding to a guanine-N7 (either random or probably part of a GG sequence), the product would result – after digestion – in the adduct $PtX(GMP)(NH_3)_2$. This species still has a reactive ligand X ($H_2O$, $OH^-$, $Cl^-$) and will not live very long, unless no nearby guanine or adenine is available. In that case the monofunctionally bound product may be stabilised, e.g. by hydrogen bonding to the O6-atom. This kind of binding in fact has been found by Wing et al.[117], in a crystal structure determination of a dodecanucleotide with alternating CG sequences. In other cases, the monofunctionally bound cisplatin will rapidly, or may be even instantaneously (see above), react with a neighboring guanine or adenine, or – in a much slower reaction – a more distant purine site. Recent findings have made clear that after very short reaction times, monofunctional adducts can indeed be detected[156, 157]. How these reactions will occur under in vivo conditions, with much smaller concentrations, will be discussed in the next section.

## 9 Reactions of Platinum Amine Compounds in Vivo

Apart from the binding of cisplatin and analogs to DNA in the tumor cell – which is generally believed to be the origin of the antineoplastic activity – several other events will take place after administration of the drug into the body (usually through infusion in the blood stream). Although other classes of drugs may be given simultaneously in combination therapy (e.g., vinblastin, bleomycin), present evidence suggests that at best weak chemical reactions occur between the drugs[17, 152]. On the other hand, a large variety of reactions may take place during the transport through the body via the blood. As discussed above, hydrolysis reactions are severely suppressed because of the high $Cl^-$ concentration in the blood. Nevertheless, reactions with e.g. blood proteins or peptides may occur, although systematic studies about the molecular basis of these reactions are not yet available. Likely candidates for binding may be the thiol and thioether groups of proteins and peptides, which are known to have a high affinity for platinum[1, 2].

During the first 24 h after the application of the drug, already 30–70% of the administered platinum is excreted from the body, mainly through the kidneys. Initially this leads to a severe problem, because of the resulting kidney damage. Nowadays, the patients receiving cisplatin are simultaneously treated with a high dose of fluid that prevents this damage[17]. Other toxic effects in the early days, when high doses of cisplatin were administered, were probably due to the formation of dinuclear (and even trinuclear) hydroxo-bridged Pt-species[16, 17, 20]. In later administration protocols, the platinum concentrations were lowered, so that only monomeric species are supposed to enter the body.

The circulating platinum species is transported – most likely as a neutral species – into both tumor cells and normal cells, and then undergoes hydrolysis inside the cells (see Fig. 6). Subsequently, the hydrolyzed Pt-species can react with all kinds of cell components, such as proteins, DNA and RNA.

Although the binding to DNA appears to be responsible for the ultimate cell killing, it cannot be excluded that reactions of the Pt species with other molecules are leading to toxic side effects. These effects can be classified as acute (gastrointestinal damage and hypersensitivity), subacute (nephrotoxicity; reproductive toxicity) and chronic (nephro- and neurotoxicity). The molecular basis of these toxicities is far from understood, although medical research has resulted in protocols that reduce the effects enormeously; (e.g. the pretreatment hydration of the patient, which almost avoids nephrotoxicity). Detailed information on the characteristics of the toxic side effects is considered to be outside the scope of this review[118, 119]. A schematic representation of the transport of cisplatin through the body is depicted in Fig. 20.

Recently, it could be established that in living cells the same types of Pt-adducts are formed as in DNA treated with cisplatin in vitro[120, 121]. Also the strong preference of cisplatin to form intrastrand crosslinks on GG base sequences is found upon interaction with DNA in vivo[120].

The nature of the difference in response to Pt-chemotherapy between tumor and normal cells is still unknown. It is possible that rapidly dividing (tumor) cells are more susceptible because of the inhibitory effect of Pt-DNA adducts during the DNA replication, a prerequisite for cell division. Another explanation may be a difference in the capacities of the various cell types to remove Pt-DNA adducts. The influence of DNA repair systems on the survival of cisplatin-treated cells has clearly been demonstrated in

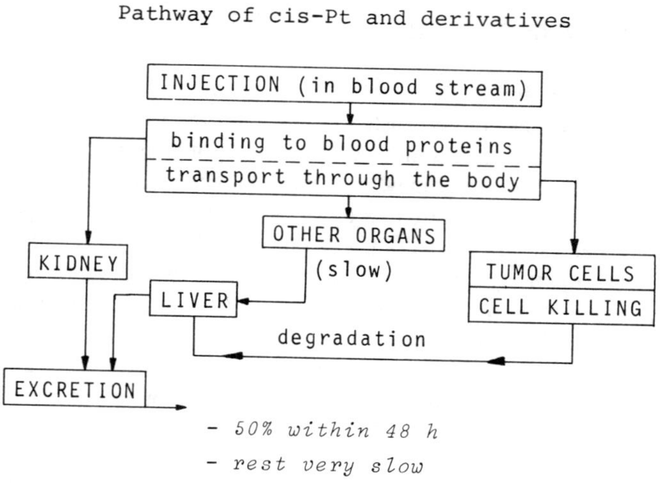


**Fig. 20.** Simplified pathway of cisplatin in the body after administration in the blood

*E. coli* bacteria[48] and in cultured human cells. In Fig. 21 the survival curves are given for human fibroblasts after treatment with various cisplatin concentrations[50]. The curves obtained for cells of patients suffering from inherited DNA repair deficiencies, such as xeroderma pigmentosum and Fanconi's anemia, clearly indicate that these cells are more susceptible to the cytotoxic activity of cisplatin than the same type of cells obtained from "normal", healthy persons.

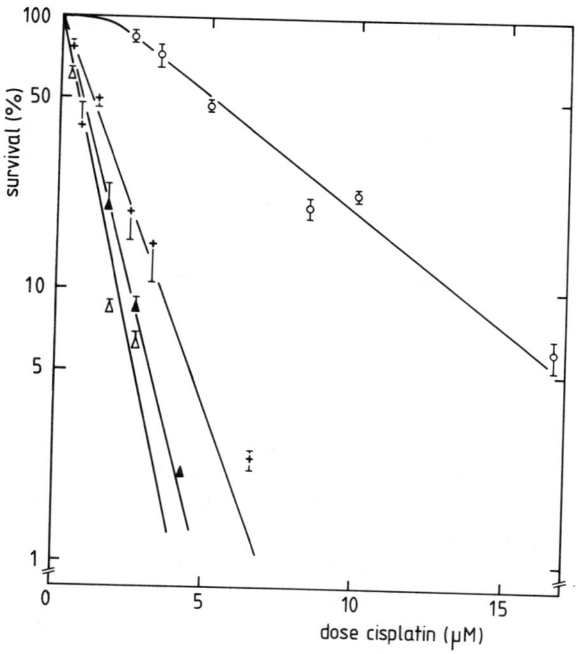


**Fig. 21.** Survival curves of human fibroblasts as a function of the dose of cisplatin used in the treatment. O = normal, healthy person; + = Fanconi's anemia; ▲ and △ = xeroderma pigmentosum, complementation group F and A, resp

The question which of the induced Pt-DNA adducts is (are) responsible for the antitumor activity is still unanswered. Many investigators have tried to correlate interstrand DNA and DNA-protein crosslink formation with the cytotoxic action of cisplatin. However, conflicting results were obtained[18, 50)]. Until now a biological role inside the cell has been indicated for only one type of Pt-DNA adduct. Brouwer et al.[48)] showed that cisplatin can induce base-pair substitutions in *E. coli* bacteria at GAG and GCG base sequences. This strongly suggests that the intrastrand crosslink of cisplatin on GBG is responsible for this effect.

Studies are ongoing to establish not only the induction but also the repair of the various Pt-DNA adducts, because it is assumed that incorrectly repaired and non-repaired, persistent, lesions will be responsible for the antitumor effects of the drug. For the study of the Pt-adduct formation and repair in DNA of cells treated with biologically relevant doses, e.g. from patients treated with cisplatin, very sensitive detection methods are required. For this purpose immunochemical techniques have been developed with which, at the moment, femtomol ($10^{-15}$ mol) amounts of the Pt-adducts in DNA can be detected with *anti*-Pt-adduct antibodies[120, 122, 123)].

Patients, suffering apparently from the same type of tumor, often respond differently to Pt-chemotherapy. Therefore, a research program has been started on the Pt-adduct formation and repair in the DNA of the nucleated peripheral blood cells of these patients. Figure 22 shows the number of intrastrand crosslinks on pGG sequences per $10^9$

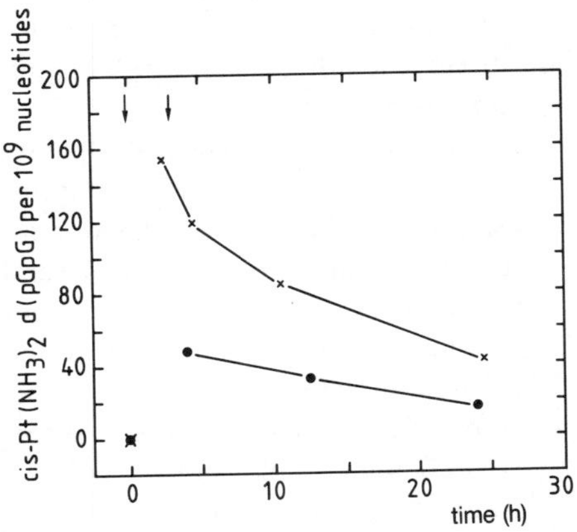


**Fig. 22.** Numbers of intrastrand crosslinks to GG-units in DNA isolated from white blood cells of two different patients at various time intervals after the onset of the cisplatin infusion (indicated by *arrows*).

nucleotides in DNA, isolated from the white blood cells of two patients after their first cisplatin infusion[120)]. In these patients unequal levels of platination have been induced immediately after the cisplatin-infusion period of 3 h. The data also show that the amounts of adducts decrease during the first 24 h after the onset of the treatment. Apparently, the main part is repaired within this period. Until now, not enough data are available to establish a possible correlation between the level of intrastrand crosslinks on GG sequences in the DNA of blood cells of patients and their response to Pt-chemotherapy, though Reed[124)] recently reported that a positive correlation between these two effects exists.

## 10 Future Developments

The progress made in the understanding of how cisplatin operates during the last decade, i.e. the period that covers the first clinical trials and the world-wide – still increasing – application of cisplatin, is enormous. Nevertheless, the number of important questions that are still unanswered remains large, and much work needs to be done by combined efforts of chemists, biochemists, biologists, pharmacologists, medical oncologists and other biomedical researchers. Topics that are likely to be important for medical application, all will have to deal with:

- the mechanism of action and side reactions in vivo of cisplatin and the origin of discrimination by the compound – if any – between a tumor cell and a normal cell
- the differences in response between the various tumor types
- the resistance of certain patients for this type of chemotherapy, already present at the onset of the treatment, or rising during the treatment procedure.

Other important questions that have to be answered are:

- Is there any discrimination by cisplatin between the cell wall of a tumor cell and a normal cell? I.e. can cisplatin recognize tumor cells?
- What is the relevance of cisplatin binding to other molecules than DNA (small and large ones) present in the cell?
- Does a molecular basis exist for synergism of cisplatin with other drugs?
- What kind of processes are involved in deplatination and what are the mechanisms leading to repair of the DNA-cisplatin binding?
- What is the effect of the binding of cisplatin on DNA processing?
- What is the influence of DNA-platination on transcription of DNA and on translation to produce proteins?

From the inorganic chemistry point of view, these questions are expected to lead to the following studies in the near future:

- Detailed studies of the kinetics of Pt-binding to oligonucleotides (with specific sequences) and to DNA (from several sources; in vitro and in vivo including cisplatin-treated patients; tumor cells, normal cells).
- Detailed studies of the structures (including crystal structures) of the formed adducts of cisplatin with oligonucleotides and DNA (again from several sources).
- Study of the molecular basis of synergism, i.e. studying the simultaneous effect upon DNA binding (and distortion) of cisplatin in combination with other drugs.
- Study of the reaction and the resulting products, of cisplatin with other cellular components than DNA.
- Studies of other Pt-compounds with respect to detailed structure determination after DNA binding.

## 11 Information Added After Completion of the Review

After the completion of this review, some exciting results have become available, which are worthwhile adding.

In the field of new platinum compounds the recently described ascorbato-Pt compounds look very promising[125, 126]. In the study of platinum compounds with nucleobases and derivatives, and mononucleotides, we would like to mention: binding of cisplatin to N1 and even N3 of purines[127–130], the observation of some head-to-head structures in the solid state[131] and the diastereotopic exchange in some bis(guanosine)Pt compounds with sterically hindered amines[132].

In the area of oligonucleotides the results of the theoretical analysis of Miller et al.[133] are of great interest, since the Pt(GG) chelate appears to be the most likely one, just as described above.

Phosphate platinum interactions have been reported by e.g. Bose[134], Appleton[135] and Marzilli[147].

From the point of view of improvements in the treatment of tumors by combination therapy, we would like to mention the review about radiation therapy in combination with cisplatin treatment[136].

Detailed studies of hydrolysis and ligand exchange kinetics[137–139] with 5′-GMP and 5′-AMP and with diadenosine tetraphosphate have illustrated that kinetics might be much more important than previously thought.

A detailed analysis of crystal structures in compounds with formula $Pt(amine)_2(purine)_2$ has made clear that H-bond interaction between the amine N–H and the O6 of the purine do occur frequently[140, 141].

It now appears that, using FTIR and Resonance Ramon spectroscopy even oligonucleotide conformations may be studied[142, 161].

Great progress was made by Marzilli[143] on using $^{1}H$-detected multiple-quantum NMR, in the structure determination of d(TGGT).

Careful analysis of $^{195}Pt$ NMR spectra in a series of Pt-nucleobase compounds has allowed binding site determinations from chemical shifts[127, 144].

Further, Lippert has reported[145] the first N3, N4 chelate of cytosine to a Pt(IV) species.

It now appears[153] that interindividual human variations in cisplatin response can be detected with an in vitro assay. Lippard et al.[159] have reported that site-specifically platinated DNA is a new probe for the biological activity of cisplatin.

*Acknowledgements.* The research summarized in this review was supported by: Leiden University; Medical Biological Laboratory TNO, Rijswijk, The Netherlands; Academic Hospital, Leiden, The Netherlands; Organisation for the Advancement of Pure Research (ZWO), the Netherlands Foundation of Chemical research (SON: grant number 11-28-17), the Netherlands Cancer Foundation KWF: granted projects MBL-79-1, MBL-83-1, IKW-83-16) and Johnson and Matthey (generous loan of platinum).

The progress of the research has benefitted very much through discussions with the research group of Prof. Dr. J. C. Chottard (Paris), which were made possible through support of the French-Dutch Cultural Agreement.

The authors are very much indebted to their many colleagues, co-workers and students for their invaluable contributions to the research described above. Most of their names appear as co-authors in the list of references.

## 12 References

1. Hartley, F. R.: The Chemistry of Platinum and Palladium, Applied Science Publishers, Essex 1973
2. Cotton, F. A., Wilkinson, G.: Advanced Inorganic Chemistry, Wiley, London 1983
3. Werner, A.: Z. Anorg. Chem. *3,* 267 (1983)

4. Peyrone, M.: J. Liebigs Ann. Chem. *51,* 15 (1845)
5. Basolo, F., Pearson, R. G.: Mechanisms of Inorganic Reactions, Wiley, New York 1967
6. Basolo, F., Chatt, J., Gray, H. B., Pearson, R. G., Shaw, B. L.: J. Chem. Soc. 2207 (1961)
7. Atwood, J. D.: Inorganic and Organometallic Reaction Mechanisms, Brooks/Cole, Montery 1985
8. Lim, M. C., Martin, R. B.: J. Inorg. Nucl. Chem. *38,* 1911 (1976)
9. Underhill, A. E., Watkins, D. M.: Chem. Soc. Rev. *9,* 429 (1980)
10. Rosenberg, B. in: Nucleic Acid Metal ion interactions, (ed. Spiro, T. G.), Wiley, New York 1980
11. Rosenberg, B., VanCamp, L., Krigas, T.: Nature *205,* 698 (1965)
12. Rosenberg, B., VanCamp, L., Grimley, E. B., Thomson, A. J.: J. Biol. Chem. *242,* 1347 (1967)
13. Rosenberg, B., VanCamp, L., Trosko, J. E., Mansour, V. H.: Nature *222,* 385 (1969)
14. Roberts, J. J., Thomson, A. J.: Progr. Nucl. Ac. Res. Mol. Biol. *22,* 71 (1979)
15. Wiltshaw, E.: Plat. Met. Rev. *23,* 90 (1979)
16. Prestayko, A. W., Crooke, S. T., Carter, S. K. (eds.): Cisplatin, current status and new developments, Academic Press, New York 1980
17. Hacker, M. P., Douple, E. B., Krakoff, I. H. (eds.): Platinum Coordination Compounds in Cancer Chemotherapy, Martinus Nijhoff, Boston 1984
18. Pinto, A. L., Lippard, S. J.: Biochem. Biophys. Acta *780,* 167 (1985)
19. Marcelis, A. T. M., Reedijk, J.: Recl. Trav. Chim. Pays-Bays *102,* 121 (1983)
20. Lippard, S. J.: Science *218,* 1075 (1982)
21. Lippert, B., Beck, W.: Chemie in unserer Zeit *17,* 190 (1983)
22. Sigel, H. (ed): Nucleotides and Derivatives; their ligating ambivalency, Metal ions in biological systems, Vol. 8, M. Dekker, New York (1979)
23. Marzilli, L. G.: Progr. Inorg. Chem. (Lippard, S. J., ed.) *23,* 255 (1977)
24. Marzilli, L. G., Eichhorn, G. L. (eds): Metal ions in Genetic Information Transfer, New York, Elsevier 1984
25. Martin, R. B.: Acc. Chem. Res. *18,* 32 (1985)
26. Cleare, M. J.: J. Clin. Hemat. Oncol. *7,* 1 (1977)
27. Kauffman, G. B.: Inorg. Synth. *7,* 249 (1963)
28. Dhara, S. C.: Indian J. Chem. *8,* 193 (1970)
29. Tobe, M. L., Khokhar, A. R.: J. Clin. Hemat. Oncol. *7,* 114 (1977)
30. Evans, B. D., Rajn, K. S., Calvert, A. H., Harland, S. J., Wiltshaw, E.: Cancer Treat. Rep. *67,* 997 (1983)
31. Van Kralingen, C. G., Reedijk, J., Spek, A. L.: Inorg. Chem. *19,* 1481 (1980)
32. Van der Veer, J. L., Hinrichs, W. L., Reedijk, J.: Acta Crystallogr. *C, 42,* 536 (1986)
33. Rochon, F. D., Melanson, R., Macquet, J. P., Belanger-Gariepy, F., Beauchamp, A. L.: Inorg. Chim. Acta *108,* 17 (1985)
34. Bruck, M. A., Bau, R., Noji, M., Inagaki, K., Kidani, Y.: ibid. *92,* 279 (1984)
35. Martin, R. B.: ACS Symposium Series (Lippard, S. J., ed.) *209,* 231 (1983)
36. Lee, K. W., Martin jr., D. S.: Inorg. Chim. Acta, *17,* 105 (1976)
37. Johnson, N. P., Hoeschele, J. P., Rahn, R. O.: Chem. Biol. Interactions *30,* 151 (1980)
38. Marcelis, A. T. M., Van Kralingen, C. G., Reedijk, J.: J. Inorg. Biochem. *13,* 213 (1980)
39. Marcelis, A. T. M., Den Hartog, J. H. J., Reedijk, J.: J. Am. Chem. Soc. *104,* 2664 (1982)
40. Marcelis, A. T. M., Erkelens, C., Reedijk, J.: Inorg. Chim. Acta *91,* 129 (1984)
41. Saudek, V., Pivcová, H., Nosková, D., Drobnik, J.: J. Inorg. Biochem. *24,* 13 (1985)
42. Dijt, F. J., Breijer, W., Kooistra, M., Reedijk, J.: unpublished
43. Vollano, J. F., Blatter, E. E., Dabrowiak, J. C.: J. Am. Chem. Soc. *106,* 2732 (1984)
44. Van der Veer, J. L., Peters, A. R., Reedijk, J.: J. Inorg. Biochem., *26,* 137 (1986)
45. Reslova, S.: Chem. Biol. Interactions *4,* 66 (1971)
46. Harder, H. C., Rosenberg, B.: Int. J. Cancer *6,* 207 (1970)
47. Howle, J. A., Gale, G. R.: Biochem. Pharmacol. *19,* 2757 (1970)
48. Brouwer, J., Van de Putte, P., Fichtinger-Schepman, A. M. J., Reedijk J.: Proc. Natl. Acad. Sc. *78,* 7010 (1981)
49. Plooy, A. C. M.: Ph.D. Thesis, State University Leiden, 1984
50. Plooy, A. C. M., Van Dijk, M., Berends, F., Lohman, P. H. M.: Cancer Res. *45,* 4178 (1985)

51. Gellert, R. W., Bau, R. in: Metal Ions in Biological Systems (Sigel, H., ed.), M. Dekker, New York, *8,* 1 (1979)
52. Hodgson, D. J.: Progr. Inorg. Chem. *23,* 211 (1977)
53. Lippard, S. J. (ed.): Platinum, Gold and other Metal Chemotherapeutic Agents: Chemistry and Biochemistry, ACS Symposium Series, *209,* Washington (1983)
54. Duncan, R. E., Tobias, R. S.: Inorg. Chem. *16,* 2625 (1977)
55. Lallemand, J. Y., Soulié, J., Chottard, J. C.: J. Chem. Soc., Chem. Comm., 436 (1980)
56. Mansy, S., Chu, G. Y. H., Duncan, R. E., Tobias, R. S.: J. Am. Chem. Soc. *100,* 607 (1978)
57. Inagaki, K., Tamaoki, N., Kidani, Y.: Inorg. Chim. Acta *46,* L93 (1980)
58. Noszal, B., Scheller-Krattiger, V., Martin, R. B.: J. Am. Chem. Soc. *104,* 1078 (1982)
59. Van der Veer, J. L., Ph.D. thesis, Leiden University (1986) and unpublished observations
60. Rosenberg, B.: Biochemie *60,* 859 (1978)
61. Cramer, R. E., Dahlstrom, P. L.: J. Am. Chem. Soc. *101,* 3679 (1979)
62. Ciccarelli, R. B., Solomon, M. J., Varshavsky, A., Lippard, S. J.: Biochemistry, *24,* 7533 (1985)
63. Sherman, S., Gibson, D., Wang, A. H. J., Lippard, S. J.: Science *230,* 412 (1985)
64. Kozelka, J., Petsko, G. A., Lippard, S. J., Quigley, G. J.: J. Am. Chem. Soc. *107,* 4079 (1985), and Inorg. Chem. *25,* 1075 (1986)
65. Roos, I. A. G., Thomson, A. J., Mansy, S.: J. Am. Chem. Soc. *96,* 6484 (1974)
66. Jordanov, J., Williams, R. J. P.: Bioinorg. Chem. *8,* 77 (1978)
67. Inagaki, K., Kidani, Y.: J. Inorg. Biochem. *11,* 39 (1979)
68. Chottard, J. C., Girault, J. P., Chottard, G., Lallemand, J. Y., Mansuy, D.: Nouv. J. de Chim. *2,* 551 (1978)
69. Marcelis, A. T. M., Canters, G. W., Reedijk, J.: Recl. Trav. Chim. Pays-Bas *100,* 391 (1981)
70. Chottard, J. C., Girault, J. P., Chottard, G., Lallemand, J. Y., Mansuy, D.: J. Am. Chem. Soc. *102,* 5565 (1980)
71. Caradonna, J. P., Lippard, S. J., Gait, M. J., Singh, M.: ibid. *104,* 5793 (1982)
72. Den Hartog, J. H. J., Altona, C., Chottard, J. C., Girault, J. P., Lallemand, J. Y., De Leeuw, F. A. A. M., Marcelis, A. T. M., Reedijk, J.: Nucl. Acids, Res. *10,* 4715 (1982)
73. Girault, J. P., Chottard, J. C., Guittet, E. R., Lallemand, J. Y., Huynh-Dinh, T., Igolen, J.: Biochem. Biophys. Res. Commun. *109,* 1157 (1982)
74. Inagaki, K., Kasuya, K., Kidani, Y.: Chem. Lett. 1345 (1983)
75. Chottard, J. C., Girault, J. P., Guittet, E. R., Lallemand, J. Y., Chottard, G.: ACS Symposium Series (Lippard, S. J., ed.), *209,* 125 (1983)
76. Marcelis, A. T. M.: Ph.D. thesis, State University Leiden (1982)
77. Inagaki, K., Kidani, Y.: Inorg. Chim. Acta *80,* 171 (1983)
78. Girault, J. P., Chottard, J. C., Neumann, J. M., Tran-Dinh, S., Huynh-Dinh, T., Igolen, J.: Nouv. J. de Chim. *8,* 7 (1984)
79. Inagaki, K., Kasuya, K., Kidani, Y.: Chem. Lett. 171 (1984)
80. Den Hartog, J. H. J., Altona, C., Van Boom, J. H., Van dere Marel, G. A., Haasnoot, C. A. G., Reedijk, J.: J. Am. Chem. Soc. *106,* 1528 (1984)
81. Den Hartog, J. H. J., Altona, C., Van Boom, J. H., Marcelis, A. T. M., Van der Marel, G. A., Rinkel, L. J., Wille-Hazeleger, G., Reedijk, J.: Eur. J. Biochem. *134,* 485 (1983)
82. Inagaki, K., Kidani, Y.: Inorg. Chim. Acta, *92,* L9 (1984)
83. Marcelis, A. T. M., Den Hartog, J. H. J., Van der Marel, G. A., Wille, G., Reedijk, J.: Eur. J. Biochem. *135,* 343 (1983)
84. Inagaki, K., Kasuya K., Kidani, Y.: Inorg. Chim. Acta *91,* L13 (1984)
85. Den Hartog, J. H. J., Altona, C., Van Boom, J. H., Reedijk, J.: Recl. Trav. Chim. Pays-Bas *103,* 322 (1984)
86. Girault, J. P., Chottard, G., Lallemand, J. Y., Huguenin, F., Chottard, J. C.: J. Am. Chem. Soc. *106,* 7227 (1984)
87. Den Hartog, J. H. J., Altona, C., Van Boom, J. H., Reedijk, J.: FEBS Letters, *176,* 393 (1984)
88. Van Hemelryck, B., Guittet, E., Chottard, G., Girault, J. P., Huynh-Dinh, T., Lallemand, J. Y., Igolen, J., Chottard, J. C.: J. Am. Chem. Soc. *106,* 3037 (1984)
89. Den Hartog, J. H. J., Altona, C., Van der Marel, G. A., Reedijk, J.: Eur. J. Biochem. *147,* 371 (1985)

90. Den Hartog, J. H. J., Altona, C., Van den Elst, H., Van der Marel, G. A., Reedijk, J.: Inorg. Chem. *24,* 986 (1985)
91. Den Hartog, J. H. J., Altona, C., Van Boom, J. H., Van der Marel, G. A., Haasnoot, C. A. G., Reedijk, J.: J. Biomol. Struc. Dyn. *2,* 1137 (1985)
92. Girault, J. P., Chottard, G., Lallemand, J. Y., Chottard, J. C.: Biochemistry *21,* 1352 (1982)
93. Altona, C.: Recl. Trav. Chim. Pays-Bas *101,* 413 (1982)
94. Den Hartog, J. H. J.: Ph.D. Thesis, State University Leiden, 1985, and unpublished observations
95. Inagaki, K., Kidani, Y.: Inorg. Chim. Acta *106,* 187 (1985)
96. Admiraal, G., Van der Veer, J. L., De Graaff, R. A. G., den Hartog, J. H. J., Reedijk, J.: J. Am. Chem. Soc. *109,* 592 (1987)
97. Brouwer, J., Fichtinger-Schepman, A. M. J., Van de Putte, P., Reedijk, J.: Cancer Res. *42,* 2416 (1982)
98. Van der Veer, J. L., Ligtvoet, G. J., Van der Elst, H., Reedijk, J.: J. Am. Chem. Soc. *108,* 3860 (1986)
99. Van der Veer, J. L., Van der Elst, H., Den Hartog, J. H. J., Reedijk, J.: Inorg. Chem. *25,* 4657 (1986)
100. Fichtinger-Schepman, A. M. J., Lohman, P. H. M., Reedijk, J.: Nucl. Acid Res. *10,* 5345 (1982)
101. Marzilli, L. G., Reily, M. D., Heyl, B. L., McMurray, C. T., Wilson, W. D., FEBS Letters, *176,* 389 (1984)
102. Chottard, J. C., Private communication
103. Stone, P., Kelman, A. D., Sinex, F. M., Bhargava, M. M., Halvarson, H. O.: J. Mol. Biol. *104,* 793 (1976)
104. Cohen, G. L., Bauer, W. R., Barton, J. K., Lippard, S. J.: Science *203,* 1014 (1979)
105. Reilly M. D., Marzilli, L. G.: J. Am. Chem. Soc. *107,* 4916 (1985)
106. Rubin, J. R., Sabat, M., Sundaralingam, M., in: Pullman, B., Jortner, J. (eds.): Nucleic Acids the Vectors of Life, Reidel 1983
107. Dewan, J. C.: J. Am. Chem. Soc. *106,* 7239 (1984)
108. Fichtinger-Schepman, A. M. J., Van der Veer, J. L., den Hartog, J. H. J., Lohman, P. H. M., Reedijk, J.: Biochemistry *24,* 707 (1985)
109. Fichtinger-Schepman, A. M. J., Van der Veer, J. L., Lohman, P. H. M., Reedijk, J.: J. Inorg. Biochem. *21,* 103 (1984)
110. Pinto, A. L., Lippard, S. J.: Proc. Natl. Acad. Sci. USA, *82,* 4616 (1985)
111. Eastman, A.: Biochemistry *22,* 3927 (1983)
112. Eastman, A.: Biochemistry *24,* 5027 (1985)
113. Johnson, N. P.: Biochem. Biophys. Acta *104,* 1394 (1982)
114. Johnson, N. P., Mazard, A. M., Escalier, J., Macquet, J. P.: J. Am. Chem. Soc. *107,* 6376 (1985)
115. Rahn, R. O.: J. Inorg. Biochem. *21,* 311 (1984)
116. Olinski, R., Walter, Z.: Z. Naturforsch. *39c,* 1057 (1984)
117. Wing, R. M., Pjura, P., Drew, H. R., Dickerson, R. E.: EMBO J., *3,* 1201 (1984)
118. Loehrer, P. J., Einhorn, L. H.: Ann. Int. Med. *100,* 704 (1984)
119. Sleijfer, D. T., Meijer, S., Mulder, N. H.: Pharm. Weekbl. (Sci) *7,* 237 (1985)
120. Fichtinger-Schepman, A. M. J., Reedijk, J., Van Oosterom, A. T., Lohman, P. H. M.: IARC, Scientific Publications, *78,* 83 (1986)
121. Plooy, A. C. M., Fichtinger-Schepman, A. M. J., Schutte, H. H., Van Dijk, M., Lohman, P. H. M.: Carcinogenesis *6,* 561 (1985)
122. Poirier, M. C., Lippard, S. J., Zwelling, L. A., Ushay, H. M., Kerrigan, D., Thill, C. C.: Proc. Natl. Acad. Sci. USA *79,* 6443 (1982)
123. Fichtinger-Schepman, A. M. J., Baan, R. A., Luiten-Schuite, A., Van Dijk, M., Lohman, P. H. M.: Chem. Biol. Interact. *55,* 275 (1985)
124. Reed, E., Ozols, R. F., Fasy, T., Yuspa, S. H., Poirier, M. C.: Proc. 4th Int. Conf. Mutagens, New York, in press (1986)
125. Hacker, M. P., Khokhar, A. R., Brown, D. B., McCormack, J. J., Krakoff, I. H.: Cancer Res. *45,* 4748 (1985)
126. Hollis, L. S., Amundsen, A. R., Stern, E. W.: J. Am. Chem. Soc. *107,* 274 (1984)
127. Miller, S. K., Marzilli, L. G.: Inorg. Chem. *24,* 2421 (1985)

128. Dijt, F. J., Canters, G. W., Den Hartog, J. H. J., Marcelis, A. T. M., Reedijk, J.: J. Am. Chem. Soc. *106,* 3644 (1984)
129. Raudaschl-Sieber, G., Marzilli, L. G., Lippert, B., Shinozuka, K.: Inorg. Chem. *24,* 989 (1985)
130. Raudaschl-Sieber, G., Schöllhorn, H., Thewalt, U., Lippert, B.: J. Am. Chem. Soc. *107,* 3591 (1985)
131. Schöllhorn, H., Raudaschl-Sieber, G., Müller, G., Thewalt, U., Lippert, B.: ibid. *107,* 5932 (1985)
132. Cramer, R., Dahlstrom, P. L.: Inorg. Chem. *24,* 3420 (1985)
133. Miller, K. J., Taylor, E. R., Basch, H., Krauss, M., Stevens, W. J.: J. Biomol. Struct. Dyn. *2,* 1157 (1985)
134. Bose, R. N., Cornelius, R. D., Viola, R. E.: Inorg. Chem. *24,* 3989 (1985); J. Am. Chem. Soc. *108,* 4403 (1986)
135. Appleton, T. G., Hall, J. R., Neale, D. W., Ralph, S. F.: Inorg. Chim. Acta *77,* L149 (1983); Inorg. Chem. *25,* 720 (1986)
136. Douple, E. B.: Plat. Met. Rev. *29,* 118 (1985)
137. Eapen, S., Green, M., Ismail, I. M.: J. Inorg. Biochem. *24,* 233 (1985)
138. Segal, E., Le Pecq, J. B.: Cancer Res. *45,* 492 (1985); Biopolymers *23,* 1623 (1984)
139. Fanchiang, Y. T.: J. Chem. Soc. Dalton Trans. 135 (1986)
140. Heyl, B. L., Shinozuka, K., Miller, S. K., Vander Veer, D., Marzilli, L. G.: Inorg. Chem. *24,* 661 (1985)
141. Orbell, J. D., Solorzano, C., Marzilli, L. G., Kistenmacher, T. J.: Inorg. Chem. *21,* 3806 (1982)
142. Okamoto, K., Behnam, V., Theophanides, T.: Inorg. Chim. Acta *108,* 237 (1985)
143. Byrd, R. A., Summers, M. F., Zon, G., Spellmeyer-Fouts, C., Marzilli, L. G.: J. Am. Chem. Soc. *108,* 504 (1986)
144. Marzilli, L. G., Hayden, Y., Reily, M. D.: Inorg. Chem. in press (1986)
145. Beyerle-Pfnür, R., Schöllhorn, H., Thewalt, U., Lippert, B.: J. Chem. Soc. Chem. Commun. 1510 (1985)
146. Köpf-Maier, P., Köpf, H.: Naturwissenschaften *73,* 239 (1986)
147. Reilly, M. D., Marzilli, L. G.: J. Am. Chem. Soc. *108,* 8299 (1986)
148. Chottard, J. C.: Proc. 5th Int. Conf. Platinum Cancer Chemotherapy, Padua, (1987) (in press)
149. Van Hemelryck, B., Girault, J. P., Chottard, G., Valadon, P., Laoui, A., Chottard, J. C.: Inorg. Chem. *26,* 787 (1987)
150. Van Hemelryck, B., Guittet, E., Chottard, G., Girault, J. P., Herman, F., Huynh-Dinh, T., Lallemand, J. Y., Igolen, J., Chottard, J. C.: Biochem. Biophys, Res. Commun. *138,* 758 (1986)
151. Roberts, J. J., Friedlos, F.: Cancer Research *47,* 31 (1987)
152. Albertini, J. P.: Garnier-Suillerot, A.: Inorg. Chem. *25,* 1216 (1986)
153. Fichtinger-Schiepman, A. M. J., Van Ŏosterom, A. T., Lohman, P. H. M., Berends, F.: Mutation Res. *190,* 59 (1987)
154. Inagaki, K., Kidani, Y.: Inorg. Chem. *25,* 1 (1986)
155. Knox, R. J., Friedlos, F., Lydall, D. A., Roberts, J. J.: Cancer Res. *46,* 1972 (1986)
156. Butour, J. L., Johnson, N. P.: Biochem. *25,* 4534 (1986)
157. Fichtinger-Schepman, A. M. J.: Proc. 5th Int. Conf. Platinum Cancer Chemotherapy (1987) (in press)
158. Reilly, M. D., Marzinilli, L. G.: J. Am. Soc. *108,* 6785 (1986)
159. Pinto, A. L., Naser, L. J., Essigmann, J. M., Lippard, S. J.: J. Am. Chem. Soc. *108,* 7405 (1986)
160. Sundquist, W. I., Ahmed, K. J., Hollis, L. S., Lippard, S. J.: Inorg. Chem. *26,* 1524 (1987)
161. Perno, J. R., Cwikel, D., Spiro, T. G.: Inorg. Chem. *26,* 400 (1987)
162. Reedijk, J.: Pure & Appl. Chem. *59,* 181 (1987)
163. Lippard, S. J.: Pure & Appl. Chem. *59,* 731 (1987)

# The Chemistry of Chelating Agents in Medical Sciences

**R. A. Bulman**

National Radiological Protection Board, Chilton, Didcot, OX11 0RQ, England

In the last few years interest in chelating agents has expanded rapidly with increases in the number and variety of chelating agents. Chelating agents are no longer used in medical sciences primarily to remove toxic metals from the body. Some of their new uses are to be found in non-invasive diagnostic medicine, where they are used to carry either $\gamma$-emitting or paramagnetic cations to specific parts of the body, and in the manipulation of nucleic acids, a process which facilitates fragmentation into designated sequences of nucleotides. As antimetabolites, chelating agents might find uses in the chemotherapy of protozoa, viruses, microorganisms and cancer.

The development of NMR for nuclei of the metallic elements will lead to the use of new chelating agents as models of metalloproteins.

Future developments in the use of chelating agents in the medical sciences will rest upon the preparation of new chelating agents. Of particular value for understanding the binding of metals by proteins will be the preparation of chelating agents containing mixed functional groups.

# 1 Introduction

The role played by metal ions in living systems is now established and in recent years it has been recognized that inorganic biochemistry is a bridge between the academic disciplines of biochemistry and inorganic chemistry. In 1975 Wood[1] made the salient point that "biochemistry is the co-ordination chemistry of living systems". This dependence is well exemplified by the observation that one third of all enzymes have a metal ion as an essential component[2]; in addition it has been reported that all most all globular proteins bind a wide range of metal ions[3]. As metal ions play such an important role in biological systems it is to be expected that the phenomenon of chelation will be studied with interest in the medical sciences. Reviews on chelating agents in medicine go back 30 years[4] and have continued to appear with fluctuating frequency[5–13].

In general these reviews have dealt with the use of chelating agents as therapeutic countermeasures to either the uptake of toxic metal ions or the build-up and accumulation of essential metal ions to levels where they become toxic. Up to about 25 years ago chelating agents were almost in the category of a panacea for all ills. The use of chelation therapy in the removal of calcium from artherosclerotic plaque has been advanced on and off for many years[14] and in spite of statements to the contrary in authoritative journals[5], the use of ethylenediaminetetraacetic acid (EDTA) to treat perhaps 400,000 people at 1,000 clinics has been reported[16]. The use of chelating agents to treat porphyric psychosis has been consistently reported for over 30 years, an application which has recently been reviewed by Peters[17]. The symptomatology may be related to an imbalance of polyvalent cations and an associated blocking effect on various metalloenzyme systems.

The current uses of chelating agents in the medical sciences are categorized below, Sect. 2, and are accompanied by a brief description of each category. In the succeeding pages, Sect. 3, the various classes of the chelating agents are discussed in detail along with the uses. This procedure has been used so that the chelating agents are discussed primarily in their chemical groups rather than by their uses. The inter-relationships of the chelating agents and their applications are cross-referenced in Table 1; a cursory examination of Table 1 will reveal that a specific application can appear under a discussion of several chelating agents.

# 2 Current Applications

## 2.1 *Mobilization of Toxic Metal Ions*

Therapy to counter the uptake of the so-called heavy metals – a classification arbitrarily assigned to metals with a density in excess of 4 to 5 – although Nieboer and Richardson[18] have put forward an alternative classification – has, in general, been directed at mobilizing arsenic, cadmium, lead, mercury and nickel (Table 2, see page 102).

In addition to the mobilization of these four elements, mobilization of aluminium and radionuclides originating from the nuclear fission process is also of concern under this classification.

**Table 1.** Cross referencing of applications of chelating agents with chemical groupings or individual chelating agents

| Applications | Chemical Grouping or Individual Chelating Agents |
|---|---|
| 2.1 | 3.1.1; 3.1.2; 3.1.4; 3.2; 3.3.2; 3.4; 3.6.2; 3.8; 3.9, DFOA |
| 2.2 | 3.3.2; 3.5.1; 3.7 |
| 2.3 | NTA |
| 2.4 | 3.1.1; 3.1.1.1; 3.1.2; 3.1.2.1; 3.1.3; 3.1.4; 3.1.5; 3.1.6; 3.2; 3.3.2; 3.3.3; 3.4.2; 3.6.1; 3.6.2; 3.8 |
| 2.5 | DFOA |
| 2.6.1 | 3.1.2.1; 3.1.6; 3.4; 3.6.1 |
| 2.6.2 | 3.8 |
| 2.7 | EDTA, DTPA, DFOA |

## 2.2 Mobilization of Essential Elements Elevated to Toxic Levels

At the moment only two elements, iron and copper, fall into this category. It can not be excluded that other diseases might be attributed to this phenomenon. The accumulation of iron at toxic levels – the average human fatal dose is 200–250 mg $kg^{-1}$ of soluble iron[19] – can be the result of the accidental uptake of iron supplement preparations by infants, or siderosis accompanying the excessive consumption of alcoholic drinks containing large quantities of iron or siderosis accompanying regular blood transfusion. This regular use of blood transfusion is a therapy employed to counter β-thalassaemia, also known as Cooley's anaemia[20]. An effective cheap therapy is an urgent requirement as the incidence of this disease is expected to increase. In 1978 it was estimated that treatment of Britain's 300 thalassaemics would cost £ 0.75 m and that this bill might double in 10 years with new births[21]. In other countries such as Italy the bill might approach £ 30 m. At the moment regular transfusion, still the only way to counter Cooley's anaemia, can result in the body acquiring around 3 g of iron per year – a quantity which the body can not clear and which builds up in the liver and the heart. Research into therapeutic countermeasures has resulted in a large number of chelating agents being synthesized. In recent years in excess of 100 metal ion-binding agents have been screened. However, only a few have passed through to the clinical stage. The chelating agents screened for ability to mobilize intracellular iron are a diverse group and range from pyridoxal isonicotinoyl hydrazone[22] and its analogues[23], through liposomally-entrapped polyaminopolycarboxylic acids[24, 25] to siderophores of microbial origin and their biomimetic counterparts, a topic discussed in Sect. 3.3.3. Noteworthy, for the variety of chelating agents studied, over 70, is the work by Pitt et al.[26] who examined natural and synthetic hydroxamic acids, phenols, catechols and tropolones all known to have a high affinity for iron(III) in vitro. This study, in which hypertransfused male mice were employed as the model of iron overload, concluded that ethylenediamine-N,N′-bis(2-hydroxyphenylacetic acid) was the most effective therapy and was superior to desferrioxamine (DFOA); in addition it was orally active. The diversity of the chelating agents studied is remarkable, as is ingenuity of the organic chemists who synthesize silicon-containing ferric-binding agents such as tetramethyl-1,2-bis(3′-[2″,3″-diacetoxyphenyl]propyl)disiloxane).

Wilson's disease, an autosomally recessively inherited disorder, is due to a defect in copper metabolism which leads to an accumulation of the element in the liver. A subsequent effect is the accumulation of copper in the brain and the accompanying neurological disorders which characterize it. A brief account of the nature of this disease – as well as other disorders of copper metabolism – and the applications of chelation therapy appears elsewhere[13]. Indian childhood cirrhosis and primary biliary cirrhosis possess characteristics which set them apart from Wilson's disease. In the former disease, its early onset and the normal caeruloplasmin characterize it. As there is a familial incidence of the disease it can not be excluded that the hepatic accumulation of copper arises from an inherited metabolic disorder. However, the use of tinned-copper utensils in some households in India might be the source of excess dietary copper. In the latter disease copper is in the cell supernatant, unlike Wilson's disease where copper is associated with lysosomes.

## *2.3 Regulation of Metal Ions at Physiologically Desirable Levels*

At the moment only Menkes' disease, a condition which arises from an impairment of copper transport across the gut wall[27], is in this category. Again, it is possible that other diseases of this nature will be identified. Suitable copper supplementation will suppress development of the disease in the foetus. The disease is characterized by progressive degeneration of the brain and the spinal cord of infants.

## *2.4 Non-Invasive Diagnostic Medicine (NIDM)*

One of the great boons which has accompanied the development of the nuclear physics and nuclear chemistry is the use of radioactive isotopes in NIDM. In general, these radioactive isotopes are $\gamma$-emitters, although a few are positron-emitters but $\gamma$-rays are generated – in this case two opposing 511 keV $\gamma$-rays are formed by electron-positron annihilation. The potential of radiotherapy using chelated $\beta$-emitting polyvalent radiocations targetted to tumours is under investigation[28].

The success of nuclear medicine arises from the detection and diagnosis of diseases without having to resort to exploratory minor surgery; in many cases diagnosis can be made in the out-patient clinic. A brief account of the use of radioactivity for in vivo imaging has been given by Neirinckx[29]. The principal isotope used in nuclear medicine is $^{99m}Tc$, $t_{1/2}$ 6.02 h, arising as a daughter product of $^{99}Mo$, $t_{1/2}$ 66 h. $^{99m}Tc$ is eluted with sterile saline from an alumina column on which the parent nuclide is immobilized. In the saline solution $^{99m}TcO_4^-$ is present and this is reduced to the lower oxidation states Tc(V), Tc(VI), Tc(III) or Tc(I), depending on the reducing agent. The chemistry of technetium and its radiopharmaceuticals has been recently recounted[30, 31].

Research into the chelation of $^{99m}Tc$ in oxidation states other than complexed Tc(VII) has been extensive. In recent years $^{99m}Tc$ complexed by bis-(1,2-dimethylphosphino)ethane (DMPE), as $[Tc(DMPE)_3]^+$, has become an alternative to $^{201}Tl^+$ as a myocardial imaging agent[32–35]. The complexation of $^{99m}TcO^{3+}$ as neutral lipid soluble chelates of 3,3′-(1,3-propandiyldiimino)bis-(-3-methyl-2-butanone)-dioxime has facilitated the crossing of the blood-brain-barrier[36]. The development of a range of techniques which could be used to monitor the metabolism of the brain is still awaited.

Radiopharmaceuticals are also formed from radionuclides of gallium(III),indium(III) and a few other polyvalent cations. The applications of these species are discussed in subsequent sections.

In recent years the use of non-radioactive chelated cations in NMR imaging has extended NIDM. The basis of NMR imaging has been discussed elsewhere[37, 38]. Chelated cations, principally gadolinium(III), as GdDTPA, and manganese(II), are employed in NMR imaging as proton contrast reagents.

## 2.5 Antiinflammatory Agents

Copper complexes are the principal antiinflammatory agents under investigation at the moment[39, 40], although elsewhere in this review one or two other chelates are also discussed. The mechanisms by which copper complexes exert antiinflammatory action are no doubt numerous. However, it has been shown that some low molecular weight complexes dismute superoxide ($O_2^-$)[41, 42].

## 2.6 Antineoplasial, Antiviral, Antimicrobial and Antiparasitic Agents

The placing of these four topics under one heading is a convenience which can be justified on the basis that the chelating agents are, with some notable exceptions, inhibiting cellular function by inactivating metalloenzymes. The exceptions are complexes of platinum which inhibit synthesis of DNA by forming intrastrand cross-links; currently an antineoplasia treatment employs cis-diamminodichloroplatinum(II), cisplatin, but this complex is in the process of being augmented by second generation platinum(II) complexes some of which contain chelated ligands (e.g. dicarboxylates). This topic is discussed further in subsequent paragraphs.

### 2.6.1 Metalloenzymes as Target Sites

Selective inactivation of the metalloenzymes of an invader cell without destruction of the healthy cells of the host has been pursued as a chemotherapeutic procedure for many years. This is obviously an extensive topic. Unfortunately there is no extensive collation of chelates or chelating agents which inactivate metalloenzymes. The recent reviews by Perrin and Stünzi[43] and by Hutchinson[44] are primarily concerned with the antiviral actions of chelating agents and chelates. Antiviral action has been reported for EDTA, 8-hydroxyquinoline(oxine), β-diketones, 1,10-phenanthroline, phosphonoacetic acid, phosphonoformic acid, flavanoids and thiosemicarbazones[43, 44]. The mode of action can vary from depletion of calcium, an essential co-factor for some neuraminidases, through inhibition of synthesis of viral DNA by α-(N)-heterocyclic thiosemicarbazones and phosphonoacetic acid, inhibition of viral RNA synthesis by 2-(α-hydroxybenzyl)-benzimidazole to the various steps in synthesis of proteins, some of which are inhibited by isatin β-thiosemicarbazone.

In many cases nucleic acid polymerases are zinc-dependent enzymes. Hutchinson et al.[45] have drawn upon the use of phosphonoacetic acid and phosphonoformic acid and introduced mono- and bis-thiopyrophosphate. They propose that the inhibition of influenza virus occurs by inhibition of RNA transcriptase[45]. By using $^{31}P$-NMR they demonstrated chelation of zinc through sulphur in bis-thiophosphate and chelation of magnesium through oxygen. The antiviral action of some thiosemicarbazones has been known for many years, a topic discussed in Sect. 3.6.1.

The carcinostatic properties of copper complexes are been reviewed and this topic is also examined in Sect. 3.6.1. Interestingly, copper does not appear to be a constituent of any known virus-specific enzymes[43)]. The contribution of copper in regulating cellular function in many cases is associated with its redox properties. It is also possible that the redox properties of vanadium could have an important role in regulating cellular function[46)]. The antitumour properties of several vanadium(V) peroxo chelates have been studied and measured by increased life-span of mice infected with L1210 murine leukaemia. Peroxo chelates formed with oxalate, citrate, malate and iminodiacetate display antitumour activity over a relatively low and narrow dose range which is followed by a rapid onset of toxicity at doses in excess of a few mg $kg^{-1}$. In contrast, the chelates formed from NTA and EDTA gave rise to a much larger tolerance of vanadium by mice; in these cases toxicity started at doses in excess of 169 mg $kg^{-1}$.

The carcinostatic activity of bis-ketopiperazine, a derivative of a homologue of EDTA is discussed in Sect. 3.1.2.1.

The antimicrobial properties of chelating agents were among the first of the biological effects studied and this has been a topic regularly updated by Albert[47)]. Investigations with oxine and analogues established that its site of action was within either the cell or the cell membrane. The inhibition of the antibacterial action of oxine by cobaltous salts points to the catalytic degradation of hydrogen peroxide.

The mode of action of oxine on fungi does not appear to lie with the production of peroxide. Investigations into the antifungal mechanism of oxine still continues. Youatt[48)] has recently proposed that copper and ferric oxinates disorganize fungal transport systems. Gershon and Shanks[49)] advance the inhibition of lipoic acid synthesis as the mode of action but point out that it is separate from the action of copper oxinate. In recent years the extensive research in the mode of action of oxinate and its analogues has been replaced by studies on 2,2′-bipyridyls and the structurally related 1,10-phenanthrolines – topics reviewed, in English, by Pijper[52)] and Smit[51)] for doctorate theses in The Netherlands. The mode of action 2,9-dimethyl-1,10-phenanthroline against *Paracoccus denitrificans* and *Mycoplasma gallisepticum* has been attributed to the ionophoretic transport of Cu(I) to receptor liquids buried inside cell membranes[52, 53)]. The anti-mycoplasmal activity of the 2,2′-bipyridyl analogues has been attributed to their uptake of copper into the cell membrane[54)]. Effective countermeasures against phytomycoplasma, particularly in tropical agriculture, could be of great economic value. 3,4,7,8-Tetramethyl-1,10-phenanthrolinate nickel(II) has been used as a disinfectant in the cleansing of new born babies[55)].

Antibacterial activity has been demonstrated for 2-pyridinethiols, used in hair shampoos[56)], pyridine-1-oxide derivatives[57)] and a wide range α-N-heterocyclic ketone thiosemicarbazones and thiosemicarbazides[58)]. Minimum inhibitory concentrations ranged from 32 μg $ml^{-1}$ down to 0.03 μg $ml^{-1}$. The potential for antimicrobial activity among the thiosemicarbazones and thiosemicarbazides is very wide. Although the pharmaceutical industry is shifting away from exploiting fermentation products for antimicrobial properties research is not yet in the direction of chelating agents. Few reviews of the exploitation of chelation in totally synthetic antimicrobial agents have appeared.

The recognition that the interaction of the iron-transporting systems of the host and pathogen determine the outcome of bacterial infection has led to some developments which exploit the iron-transporting system. Enterobactin, a cyclic trilactone of 2,3-dihydroxy-N-benzoyl-L-serine, is secreted by a number of enterobacterial pathogens. This molecule is highly selective for ferric ions and can pull iron from transferrin and lactoferrin. Rogers[59–61] and his associates have shown that scandium and indium chelates of enterobactin act as antimetabolites to the ferric complex in *Klebsiella pneumoniae* and *Escherichia coli*. Determination of competitive uptake of the enterobactinates showed scandium, gallium and aluminium achieved 20%, 3% and 0%, respectively, of the ferric chelate. Intracellular degradation of the scandium chelate produced free scandium(III) which inhibited macromolecular synthesis in the order: RNA, protein, DNA and phospholipid[61].

The molluscidal action of 5,2′-dichloro-4′-nitrosalicylanilide has been attributed to the chelation of iron(III)[62]. The importance of chelation of iron as an antiprotozal therapy is emphasized by the suppression of experimentally induced *Plasmodium vinckei* infections in mice[63].

### *2.6.2 Nucleic Acids as Target Sites*

The foundation of metal ion and nucleic acid interactions as a chemotherapy originate with the well known work of Rosenberg[64] on platinum(II) complexes. Those containing chelating agents could destabilize nucleic acid structures by removing cations which promote the stability of histones.

Investigations of the potential of inorganic complexes as inhibitors of neoplasia are becoming extensive. In 1982 Sadler[65] reported the US National Cancer Institute as having data on file for 11,000 metal and metalloid compounds of 55 elements. Out of these 55 elements at least 35 had one active compound and for 10 elements at least 10% of the test compounds displayed some activity. Of course these active compounds need not necessarily be chelates. The antitumour properties of metal complexes have been reviewed[66] while Rosenberg[64] has reviewed the clinical aspects of platinum anticancer drugs. An extensive review by Roberts and Thompson[67] indicates that the principal target site for neutral platinum complexes is the impairment of DNA replication.

The newer clinical developments in cancer chemotherapy by platinum(II) now embrace the chelates cyclobutane-1,1-dicarboxylatodiammine-Pt(II), and diammine ethylmalonato-Pt(II)[66]. Tonew et al.[68–69] have screened several aminopyridine complexes of palladium for activity against a range of enveloped DNA and RNA viruses. No activity was found against naked RNA viruses. 4-Methyl-2-amino Pd(IV) chloride showed activity against a similar range of enveloped DNA and RNA viruses and was also active against a naked DNA virus[69].

## *2.7 Metal Chelation and Oxygen Radicals*

As three recent accounts[70–72] of the manifestations of oxygen toxicity through oxygen radicals have stressed the role of metal chelation therapy in limiting tissue damage, only a brief outline of damage to tissue by oxygen radicals will be presented.

Forms of oxygen-derived species produced by activated phagocytes can cause considerable tissue damage unless limited by copper-zinc superoxide dismutase (SOD) which destroys superoxide, $O_2^-$, or catalase which destroys $H_2O_2$. However, at least two, if not three, other reactive forms of oxygen need to be considered[70, 71].

(i) Evidence for the formation of singlet $O_2$ in phagocytic systems is still awaited, although it has been identified in chloroplasts and the lens and retina of the mammalian eye – the pertinent references are cited elsewhere[73].

(ii) The $HO_2^{\cdot}$ radical is a more powerful oxidizing species than $O_2^-$, its formation being favoured by low pH, $< pH\,5$, a condition which can be met in some macrophages.

(iii) The hydroxyl radical, $HO^{\cdot}$, a species now under intensive scrutiny because of its tissue damaging properties, is generated by homolytic fission of the peroxide band in $H_2O_2$. In addition to generation by heat or ionizing radiation, the species is generated by the catalytic action of transition metal ions; particularly effective are iron ions complexed by adenosine or guanosine triphosphates, citrate and DNA. Some of the reaction sequences are:

$$Fe^{2+} + H_2O_2 \rightarrow Fe^{3+} + HO^{\cdot} + OH^-$$

$$Fe^{3+} + H_2O_2 \rightarrow Fe^{2+} + O_2^- + H^+$$

$$HO^{\cdot} + H_2O_2 \rightarrow H_2O + H^+ + O_2^-$$

$$O_2^- + Fe^{3+} \rightarrow Fe^{2+} + O_2$$

$$HO^{\cdot} + Fe^{2+} \rightarrow Fe^{3+} + OH^-$$

$$Cu^+ + H_2O_2 \rightarrow Cu^{2+} + HO^{\cdot} + OH^-$$

The removal of metals from sensitive sites by chelation can serve to suppress $HO^{\cdot}$ degradation of tissue. Whereas EDTA reacts with $O_2^-$ and $H_2O_2$[73], both diethylenetriaminepentaacetic acid (DTPA) and bathophenanthroline induce a diminuition in iron-dependent production of $HO^{\cdot}$; however DFOA has proved to be the most effective[76]. (The structure of DFOA is given in Sect. 3.3.2.)

DFOA has been examined for ability to inhibit chronic inflammation. In a preliminary study with rheumatoid patients, intramuscular DFOA (1 g) produced an abrupt rise in haemoglobin, a response which was taken to indicate suppression of inflammation[77]. Under this regime no ill effects were observed but this was not the case where rheumatoid patients received 3 g $d^{-1}$ for 5 days each week over 1 to 3 weeks[78]. Out of seven patients ocular abnormalities were induced in four of them but subsided on termination of DFOA treatment. Further complications of DFOA therapy were observed with two patients who lapsed into unconsciousness for up to 72 hours when they were placed upon a course of treatment with phenothiazine, a treatment initiated to suppress DFOA-induced nausea.

The studies listed above all point to the need for further investigations into ferric chelation and the suppression of oxygen radical reactions. The observation that grafted islet cells in mice undergo no rejection in animals raises the possibility that ferric selective chelating agents might have a role in transplantation surgery[79].

# 3 Chelating Agents

## 3.1 Hard Bases

### 3.1.1 Polyaminopolycarboxylic Acids

The chemistry of chelation of the polyaminopolycarboxylic acids has been discussed on several occasions and no extensive discussion can be justified here. There is great diversity in the structures, ranging from the simplest nitrilotriacetic acid, (NTA), through to some of the recently synthesized ones. However, as some of the more diverse structures prepared several years ago might be unknown to the reader they are illustrated below. These chemicals are EDTA (**1**), N-hydroxyethylethylenediamine-triacetic acid (**2**, HEDTA), 2-hydroxypropane-1, 3-diaminetetraacetic acid (**3**, HPDTA), propane-1,2-diaminetetraacetic acid (**4**, PDTA), 1,2-diaminocyclohexanetetraacetic acid (**5**, DCTA), tris-(2-aminoethyl)aminehexaacetic acid (**6**, TAAHA), cyclopentane-1,2-diaminotetraacetic acid (**7**, CPDTA), DTPA (**8**, Z = 1, X = $NCH_2 \cdot COOH$), 2:2′-bis[di(carboxymethyl)amino]diethylether (**8**, Z = 1, X = 0, BADE), 2,2′-bis[di(carboxymethyl)amino]diethylsulphide (**8**, Z = 1, X = S, BADS), 1,2-bis[2-di(carboxymethyl)-aminoethoxy]ethane-tetraacetic acid (**8**, Z = 2, X = 0, BAETA), 1,2-bis[2-di(carboxymethyl)aminoethylthio]ethane tetraacetic acid (**8**, Z = 2, X = S), triethylenetetramine-hexaacetic acid (Z = 2, X = $CH_2COOH$, TTHA). Soft acid characteristics have been introduced as in mercaptoethyliminodiacetic acid which can exist as the dimer cystaminetetraacetic acid (**8**, Z = 1, X = S–S).

In recent years the organic chemistry of the polyaminopolycarboxylic acids has been studied extensively. In passing, it should be noted that the polyaminopolycarboxylic acids were mistakenly believed to be devoid of any organic chemistry. The chemistry of derivatization is discussed in Sect. 3.1.2.

*NTA*

Just over 100 years elapsed between the synthesis of NTA and the beginnings of alarm over its presence in the environment. Initially concern over eutrophication arising from phosphate-based detergents led to extensive use of NTA as a detergent builder and the conjecture that there would be a build-up of it in the environment. In recent years this concern has led to NTA having perhaps the most comprehensive data base for any synthetic organic chemical that has widespread application in consumer products[80)]. This extensive research indicates that concern over the levels of NTA is unnecessary. Ingestion of large quantities of NTA brings about increases in zinc and manganese in some species[81, 82)]. Other than a slight increase in clearance of magnesium, NTA has no effect upon clearance of endogenous metal ions from mice[83)].

NTA has found new applications in the medical sciences: liposomally entrapped $^{111}$In-NTA has been screened for potential in detection of tumours in mice[84)] and ferric NTA has been used to stimulate in vivo lipid peroxidation.

Keen et al.[86)] have demonstrated the prophylactic value of copper(II) chelated by NTA for Menkes' disease by using as a laboratory model the crinkled mutant mouse. As this syndrome does not respond to dietary supplementation with either copper sulphate or copper acetate then therapy with chelated copper would appear to have advantages.

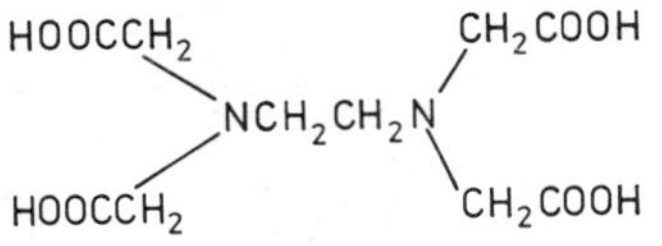


**1**

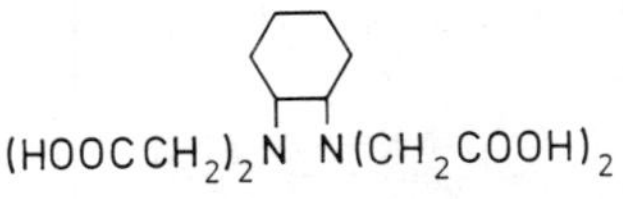


**5**

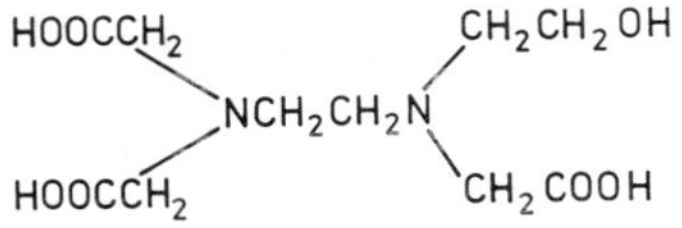


**2**

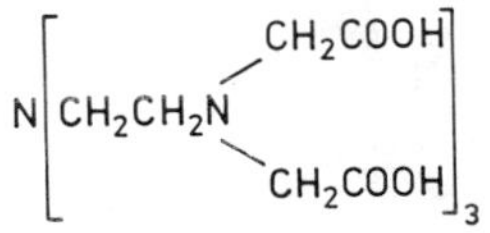


**6**

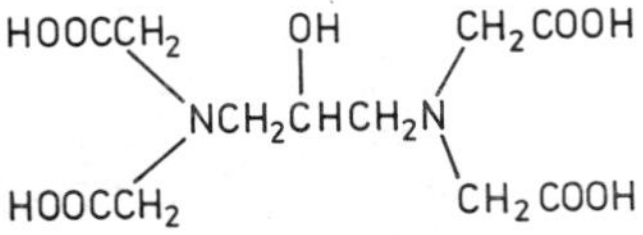


**3**

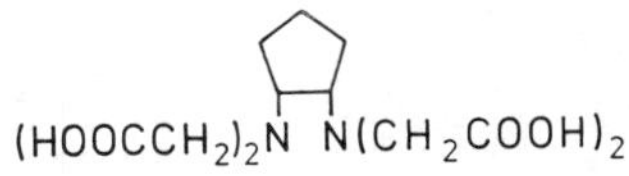


**7**

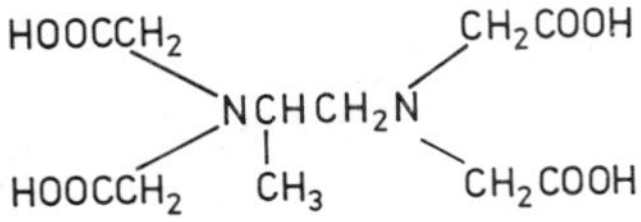


**4**

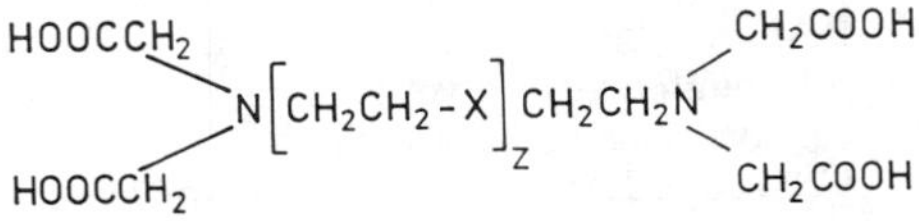


**8**

Dietary NTA has also been shown to increase the iron status of lactating mice and their pups[87].

*EDTA*

EDTA, first synthesized in 1935, was first used as a calcium-sequestering agent in the textile industry. However, in the mid-1940s EDTA rapidly became an intensely investigated substance. In addition to being examined for ability to clear toxic metals from the body – see Table 2 – it was also used to treat several other conditions. These other applications, however, are now known to be of little or no value.

As EDTA forms stable complexes with a number of physiologically important cations – $Ca^{2+}$, $Mn^{2+}$, $Cu^{2+}$, $Zn^{2+}$, $Fe^{2+}$ and $Fe^{3+}$ – it might be expected to mobilize these cations when used to treat toxic metal poisoning. Several studies have established that there is a mobilization of manganese and zinc – summarized in brief by Brownie and Aronson[88]. It is possible that the loss of zinc and manganese affects the synthesis of DNA[89]. Depletion of zinc as a consequence of the ingestion of Ca EDTA by pregnant rats results in impaired reproduction and congenital malformation[90]. In computer simulation models EDTA in plasma exists mainly as the calcium chelate (85%) whereas the zinc chelate represents only 14%[91].

In the USA there is opposition to chelation therapy, an opposition which is based upon the potential for damage to the kidneys[92, 93].

**Table 2.** Application of chelating agents to remove toxic metal ions[a]

| Element | Chelating agents[b] | | | | | | | | | |
|---|---|---|---|---|---|---|---|---|---|---|
| | BAL | PEN | AcPEN | TRIEN | DTPA | EDTA | DFOA | DMPS | DMSA | DDTC |
| As(III) | + | | | | | | | | | |
| Au(I) | + | | | | | | | | | |
| Hg(inorg)(II) | + | | + | | | | | +[e] | +[e] | |
| Hg(methyl)(II) | [c] | | | | | | | +[e] | +[e] | |
| Cd(II) | [d] | | | | | | | | | |
| Cu(II) | | + | | + | | | | | | |
| Pb(II) | | + | | | | + | | | | |
| Fe(III) | | | | | + | | + | | | |
| Ni(II) | | | | | | | | | | + |
| $^{239}$Pu(IV) | | | | | + | | | | | |
| $^{241}$Am(III) | | | | | + | | | | | |
| Al(III) | | | | | | | + | | | |

[a] Inclusion in this table does not mean that they are included national pharmacoepia, e.g., DFOA has only been used in a small number of cases to deplete the body of Al

[b] BAL, 2,3-dimercaptopropanol; PEN, D-penicillamine; AcPEN, N-acetylpenicillamine; TRIEN, triethylenetetraamine; DMPS, 2,3-dimercaptopropane sulphonic acid; DMSA, 2,3-dimercaptosuccinic acid; DDTC, N,N-diethyldithiocarbamic acid

[c] BAL may be hazardous here as it could mobilize $CH_3Hg^+$ to the brain

[d] No counteragent is yet recommended; EDTA and PEN have been proposed but they could aggravate the nephrotoxicity of Cd

[e] USSR and Eastern European Pharmacoepia only

In addition to being used to protect against toxic elements in cationic forms, EDTA in an ointment together with ascorbic acid gives protection against chromate-induced skin ulceration[94)].

*DTPA*

DTPA is superior to EDTA as an agent for removing many of the toxic metals. This superiority does not originate entirely from the greater hard base characteristics of DTPA which arise from the additional acetate moiety. DTPA has been replaced by DFOA as the treatment for iron-overload conditions. In the USA calcium and zinc salts are both approved for human use and are only available for countering uptake of transuranics[95)]. The superiority of DTPA over EDTA for polyvalent cations such as the lanthanides and the actinides arises from the potential octadenticity of DTPA as opposed to the hexadenticity of EDTA. The successful use of DTPA in reducing the $^{241}$Am burden in a worker who became peppered with $^{241}$Am-contaminated fragments in an incident in the USA in 1976 testifies to the value of chelation therapy[96)]. Intensive therapy prevented 99% of $^{241}$Am which entered the bloodstream from becoming deposited in the internal organs.

The manifestations of the toxicity of DTPA arise from depletion of transition elements[97)]. Administration of DTPA as its zinc salt suppresses its toxicity. Computer simulation studies of the chelates formed in plasma identify Zn-DTPA as the major form, 96%, whereas Ca-DTPA accounts for 3%[91)].

The high hydrophilicity and corresponding low lipophilicity of both EDTA and DTPA result in very low uptake of these chelating agents into cells. In an endeavour to increase intracellular uptake of DTPA, Rahman et al.[98] injected liposomally entrapped DTPA into mice which had hepatic accumulations of $^{239}$Pu. Although this was an effective treatment for mice it proved to be unsatisfactory for hamsters[99].

*DCTA, HEDTA and TTHA*

As these chelating agents became available 25–30 years ago they were met with an acclaim which was perhaps over enthusiastic. HEDTA as a decorporation agent of iron is inferior to DFOA and a recommendation that TTHA be used as an orally-acting therapy for the transuranics has never become practice.

*Ethylenebis[(O-hydroxyphenyl)glycine], EHPG*

EHPG, originally synthesized many years ago, has been used to model the metal coordination sites of transferrin[100]. The ester derivatives of EHPG have proved to be effective in test animals for the reduction of iron overload[101]. On the basis measurements of formation constants, Rajan et al.[102] have suggested that EHPG could be used to mobilize aluminium in aluminium-induced neurobiological disorders. In NIDM, Fe(III)-EHPG has been screened as an hepatobiliary contrast agent for the liver[103].

### 3.1.1.1 New Polyaminopolycarboxylic Acids

The general interests in chelation of metals have continued to lead to new forms of polyaminopolycarboxylic acids. Examination of the acid dissociation constant of ethylene-bis-N,N′-(2,6-dicarboxy)piperidine **(9)** showed that they were nearly identical with those of EDTA. However, the formation constants for metal chelates of **(9)** are lower by 0.5 to 3.7 log β units.

Interest in the coupling of polyaminopolycarboxylic acids to macromolecules for use in nuclear medicine resulted in the synthesis of 1-(p-aminophenyl)-ethylenediaminetetraacetic acid (**10,** n = 0, R=$NH_2$, pAPhEDTA). Since its first synthesis there have been improvements in its synthesis[105] as well as the synthesis of related chelating agents (**10,** n = 2)[106], **(11)**[107] and **(12)**[108]. In a search for a lipophilic chelator which could be incorporated into the lipid phase of liposomes, Bard et al.[109] synthesized 3-cholesteryl 6-[N′-iminobis(ethylenenitrilo)-tetraacetic acid]hexylether **(13).**

### 3.1.1.2 Polyaminopolycarboxylic Acids as Intracellular Chelators for Determining [$M^{n+}$]

Calcium is widely recognized as a major regulator of intracellular metabolism throughout the animal kingdom. Increases in intracellular concentrations of free $Ca^{2+}$ increases the activities of many enzyme systems and specialized cellular processes such as muscle contraction, cell mobility and hormone secretory processes. In the last decade or so the literature on intracellular calcium has become extensive – summarized in 1982 by Campbell[110]. As knowledge of the inorganic biochemistry of calcium has increased, improvements in the procedures for measuring intracellular free calcium have been sought.

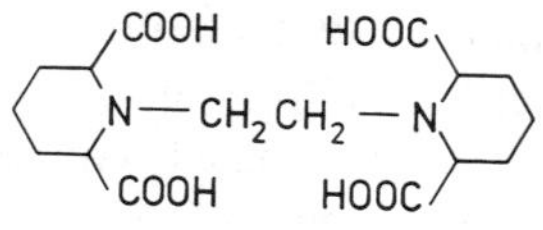

**9**

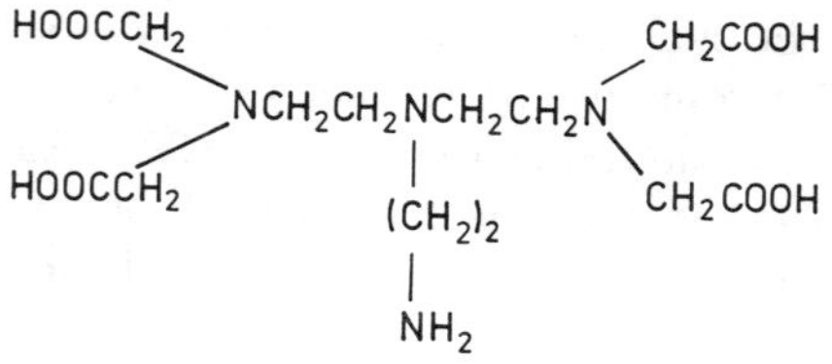

**12**

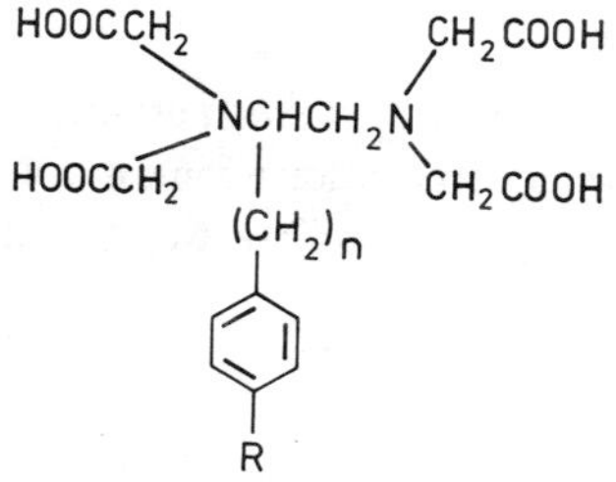

**10**

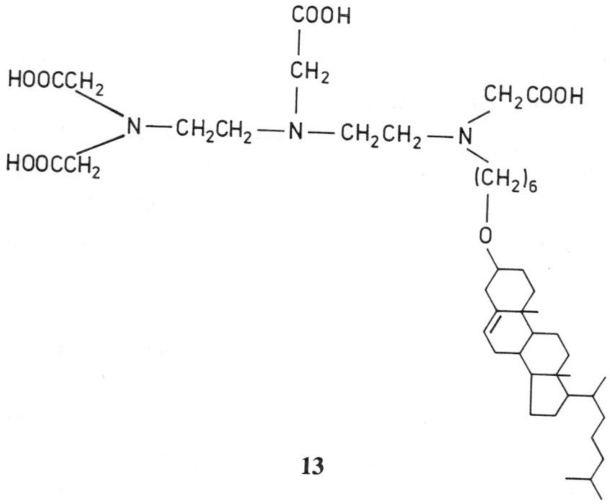

**13**

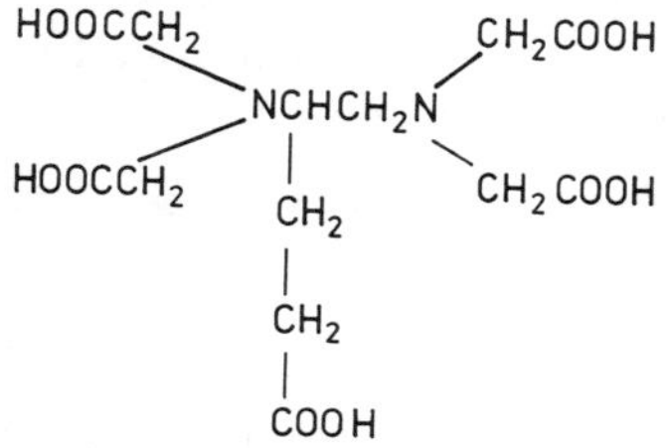

**11**

Intracellular [$Ca^{2+}$] are of the order of $10^{-6}$ M whereas extracellular levels are around $10^{-3}$ M.

In 1981 Tsien[111] introduced a procedure for loading the $Ca^{2+}$-selective chelator (**14**, R=$CH_3$, R=H) into the cytoplasm of cells. This was achieved by esterifying carboxylic groups to form the acetoxymethyl tetraester, which undergoes rapid hydrolysis in the cell to give the free chelating agent which fluoresces when complexed with $Ca^{2+}$.

In a subsequent study Smith et al.[112] incorporated fluorine into both aromatic rings (**14**, FBAPTA, R=H, $R_1$=F). The introduction of $^{19}F$ into the ring facilitates the determination of [$M^{n+}$] by measurement of $^{19}F$ chemical shifts. The [$Ca^{2+}$] measured with 5FBAPTA is unaffected by free $Mg^{2+}$ at $<10$ mM, by pH 6–8, or by minor cellular cations – $Zn^{2+}$, $Cu^{2+}$, $Mn^{2+}$, $Fe^{2+}$, species which all have high affinity for 5FBAPTA and which give separate resonances but do not affect the relative areas of the Ca-5FBAPTA and free 5FBAPTA resonances. There is a limit of about 10 ms to the response time for [$Ca^{2+}$] changes. Further developments from the same group led to quene 1 (**15**, R=$CH_3O$) and Fquene (**15**, R=F)[113]. Fquene can be used both as a fluorescent indicator and an $^{19}F$-NMR indicator and also for $pH_i$ measurements and [$Ca^{2+}$] determination in the same cell samples.

The applications of these chelating agents in measuring [$Ca^{2+}$] in cells and perfused rat heart have been summarized by Metcalfe et al.[114]. The developments of an indicator for measuring intracellular [$Na^+$] have also been reported by this group – see Sect. 3.4.1.

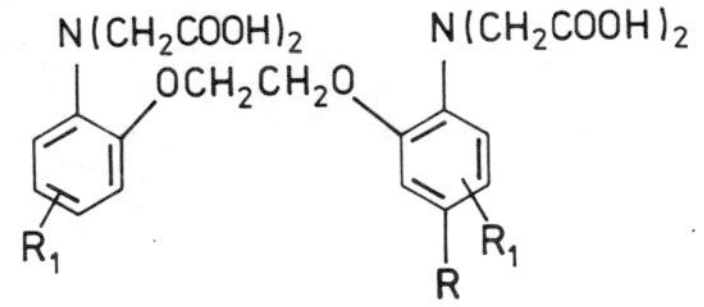


**14**

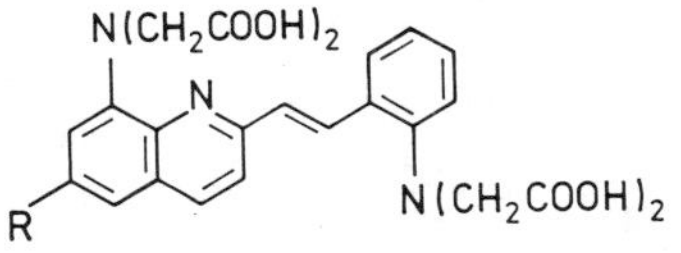


**15**

### *3.1.2 Derivatization of Polyaminopolycarboxylic Acids*

For many years the chemistry of the polyaminopolycarboxylic acids was dominated by their analytical uses and by investigations of the chelates formed with transition elements. The organic chemistry, other than the various basic syntheses, was neglected. In 1956 it was reported that these compounds were more or less totally lacking in organic chemistry[115)]. This observation, which was based upon the failure to produce anhydrides and acid chlorides, is now known to be incorrect. Although all attempts to produce acid chlorides have failed – even the action of oxalyl chloride on the anhydrides, an established procedure, was ineffective – the preparation of bisanhydride is now an established route to a variety of derivatives[116)]. In addition to the bis-anhydride, mixed acid anhydrides are available by action of acid chlorides on the triethylamine salt[116)] and water-soluble carbodiimides can be used to form amides[117)]. Other derivatives which can be formed are hydrazides[118)] and esters[119)].

Mono- and di-esters can be prepared by the action of alcohols on the bisanhydrides. The action of lithium aluminium hydride in tetrahydrofuran at $-70\,^{\circ}C$ on the tetraethyl ester of EDTA produces aldehydes in low yield[116)]. Both EDTA and DTPA as the free acids are soluble in dimethyl sulphoxide but only the bisanhydrides are soluble in dimethylformamide[116)].

The first organic derivative of DTPA to be screened for medicinal application was the pentaethyl ester which was examined for ability to reduce hepatic deposits of plutonium in mice[120)]. Although superior to DTPA in removing plutonium from mice, this derivative was too toxic for further use.

Other lipophilic derivatives of EDTA and DTPA were synthesized by condensing the bisanhydride of the former with phosphatidylethanolamine to give PE-EDTA (**16**)[121)] whereas that of the latter was condensed with 11-aminoundecanoic acid to give Puchel, **17**,[122)]. PE-EDTA failed to mobilize plutonium from the hamster, although studies with PE-[$^{14}C$]EDTA established the uptake of the radiolabel into the liver.

In contrast to PE-EDTA, Puchel mobilized plutonium from the liver of Syrian hamsters but was ineffective in mobilizing thorium from the rat[123)]. Studies with tritium-labelled Puchel in the hamster, the rat and the rabbit showed different rates of clearance. In the rat and the rabbit – species which possess a gall-bladder in contrast to the hamster – there was a rapid clearance of the chelating agent through the liver into the small intestine[124)]. In the hamster the chelating agent accumulated in the liver before being discharged into the small intestine. It is possible that these differences in hepatic clearance influence the mobilization of plutonium from the livers of hamsters and rats.

Derivatization of Puchel and EDTA led to a Puchel-$Cys_2$ (**18**)[116)] and EDTA-$Cys_2$ (**19**)[116)]. Examination of **18** and **19** for ability to mobilize cadmium from male rats showed

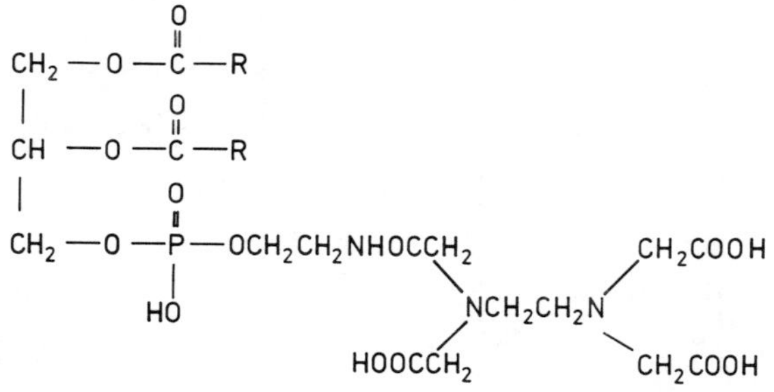

**16**

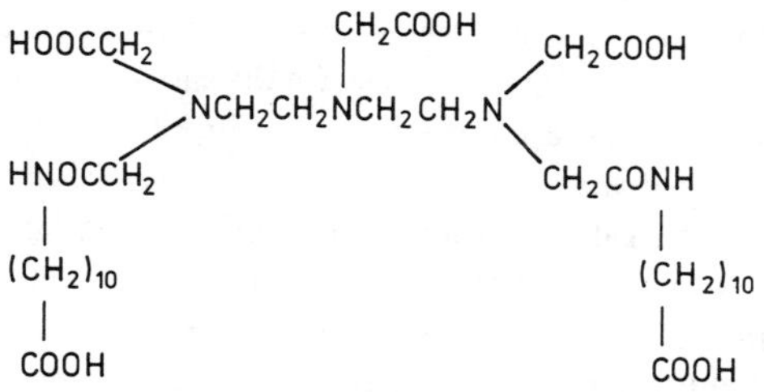

**17**

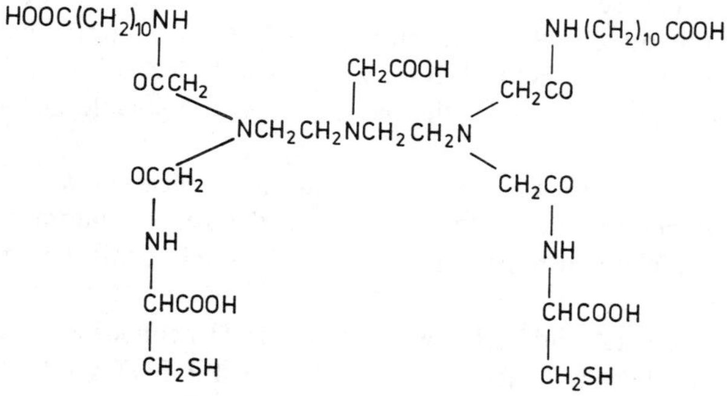

**18**

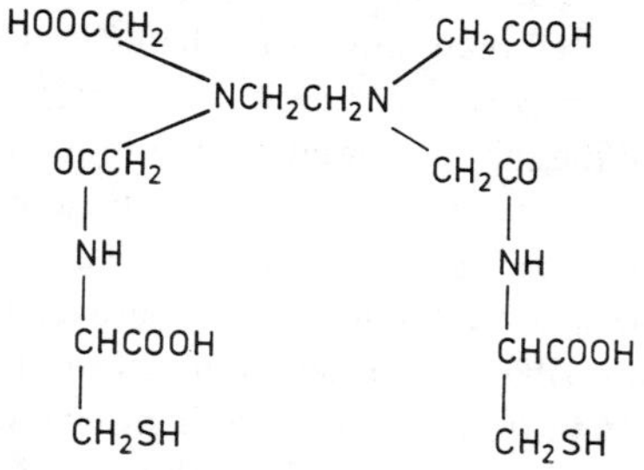

**19**

that **(19)** was ineffective whereas **(18)** reduced the body burden of cadmium by 37% when injected after a delay of 3 h. Gel chromatography of the liver cytosol of rats treated with **(18)** demonstrated that intracellular cadmium was cleared until it became associated with metallothionein[125].

Further derivatization of DTPA bis-anhydride has led to the introduction of ferric-selective moieties **(20)**[116]. Other derivatives of DTPA produced are DTPA – bis(-3-amido-4-thiol-benzotrifluoride), a chelating agent which can be monitored by $^{19}$F-NMR[116], and a fluorescent DTPA derivative formed by reacting the bisanhydride with 7-amino-4-methyl coumarin[116]. Reaction of ethylene sulphide with HPDTA leads to the 0-2-ethanethiol derivative **(21)**[116].

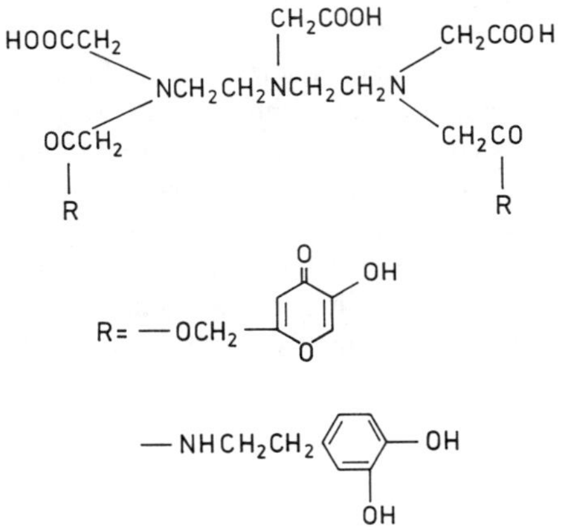


**20**

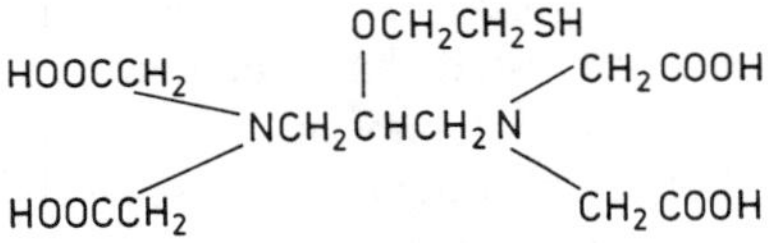


**21**

Chelating agents such as EDTA-Cys$_2$ could be of value in studying complexation of cations by metalloproteins. Some success in this area has already been achieved in modelling the complexation of cadmium by metallothionein[126]. The association of $^{113}$Cd with metallothionein, a cysteine-rich protein, has been shown by $^{113}$Cd NMR to result in $^{113}$Cd-$^{113}$Cd spin-spin coupling[127]. In the Cd-EDTA-cys$_2$ complexes spin-spin couplings and chemical shifts have been observed which are similar to those of Cd metallothionein[126]. The $^{113}$Cd-NMR of cadmium EDTA-Cys$_2$ complexes presumably arise because the remnant of the EDTA skeleton slows down $^{113}$Cd exchange rates.

### 3.1.2.1 Bisdioxopiperazines, Antitumour Agents Derived from Polyaminopolycarboxylic Acids

Suggestions by Furst[128] several years ago that many antitumour agents were either actual or potential chelating agents initiated a programme of derivatization of EDTA and its analogues at the Imperial Cancer Research Fund[129]. From this work was developed (±)-1,2-bis(3,5-dioxopiperazin-1-yl)propane (**22,** R = $CH_3$; $R_1$ = H, ICRF 159) now available in the UK as the antitumour agent Razoxane[130]. Investigations of the structure-activity relationships established the high potency of ICRF 154 (**22**, R = $R_1$ = H) in screening trials against L1210 and sarcoma 180 but no activity against experimental tumours. In contrast, ICRF 159 (**22**, R = $CH_3$; $R_1$ = H) a slightly more lipophilic chemical, in preclinical studies exhibited antitumour activity. No cytotoxic activity could be found for ICRF 192 (**22**, R = $C_2H_5$; $R_1$ = H).

Investigations of the interaction of the divalent cations of magnesium, calcium, manganese, iron, nickel, copper and zinc with the diacid diamide hydrolysis products of both ICRF 159 and ICRF 192 showed that the metal-binding affinities were similar in all cases with the exception of zinc which was bound less avidly by the hydrolysis product of ICRF 159[131)]. No cytotoxic mechanism for the zinc chelate of the hydrolysis product of ICRF 159 has been established. It is unlikely that the intracellular chelation of zinc is the primary mechanism of cytotoxicity as the hydrolysis product of ICRF 192 binds zinc most avidly.

The effectiveness of ICRF 159 had led to several other studies. In the People's Republic of China a high therapeutic index has been claimed for bimolane (**22**, R = H; $R_1$ = N-methylmorpholine) and this has led to extensive testing in the USA[132)]. X-ray diffraction studies of crystals obtained from solutions which initially contained bimolane have demonstrated the formation of ICRF 154 (**22**, R = $R_1$ = H). Camerman et al.[132)] have proposed that bimolane is inherently unstable and that the antitumour activity ascribed to it may be attributable to ICRF-154. It is possible that the N-methylmorpholine substituents increase the lipophilicity sufficiently for the basic ICRF-154 unit to be taken up by tumour cells.

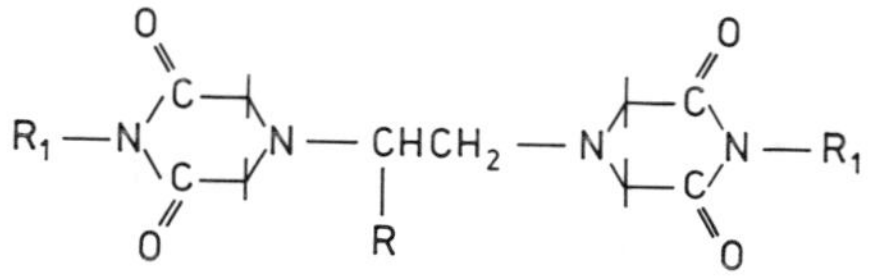

**22**

### 3.1.2.2 Polyaminopolycarboxylic Acids in NIDM

EDTA and DTPA were among the first chelating agents screened in nuclear medicine. However, their current use is rather limited with the development of chelating agents designed for selective uptake into specified regions of the body. The extensive excretion of DTPA by glomerular filtration makes it an ideal agent, for instance radiolabelled with $^{113m}In$[133)] for determining the glomerular filtration rate. When immobilized on macromolecules DTPA is also a suitable chelating agent for use in NIDM – see Sect. 3.1.2.2.

As fatty acids are taken up into the myocardium, Karesh et al.[134)] incorporated similar structures into polyaminopolycarboxylic acids, radiolabelled them with $^{99m}Tc$ and $^{57}Co$ and examined them for myocardial imaging. These chelators, which possessed chelating moieties similar to DTPA, had a low uptake into the heart ($<0.94\%$) but showed an uptake into the liver of 16%.

In a variation of the acidic function of a chelating agent, Làznickova et al.[135)] compared the biodistribution of $^{99m}Tc$ chelates of EDTA and ethylenediaminetetraacetohydroxamic acid, EDTAHA. With the exception of the heart and muscle tissue, $^{99m}Tc$-

EDTAHA had a distribution similar to $^{99m}$Tc-EDTA. Whereas $^{99m}$Tc-EDTA attained only 24% distribution in total muscle at 5 min, $^{99m}$Tc-EDTAHA attained 38% uptake. In the heart tissue the respective values were 0.6% and 0.9%.

The use of DTPA in NIDM has not been restricted to nuclear medicine. The distribution of chelates in areas of high vascularity or expanded interstitial space has been exploited in magnetic resonance studies. Wesley et al.[136)] have used gadolinium(III) – DTPA to study the magnetic relaxation times of normal and infarcted myocardium; Carr et al.[137)] and Felix et al.[138)] have used it to study cerebral tumours and Bydder et al.[139, 140)] have used it study both cerebral and cervical intraspinal tumours. Its potential for detection of human breast carcinoma has been noted[141)]. As an alternative to Gd-DTPA, Cr-EDTA, it is well tolerated by dogs and rats, could be used[142)]. As an alternative to chelators labelled with paramagnetic cations, spin-labelling with bis-piperdinyl nitroxyl derivates of EDTA and DTPA has been examined[143)].

#### 3.1.2.3 Intercalation of EDTA Derivatives into Nucleic Acid

Dougherty and Pilbrow[144)] have given a background account of physico-chemical probes of intercalation. The single-strand cleaving of DNA by oxygen and metal complexes such as ferrous bleomycin[145)], 1,10-phenanthrolinate-copper(II)[146, 147)] and cuprous streptonigrin has led to the synthesis of several DNA-binding molecules which carry ferrous EDTA[148–152)]. Hertzberg and Dervan[148)] achieved single and some double-stranded breaks in double helical DNA by coupling EDTA via 1,3-diaminopropane to methidium. In the presence of oxygen and reducing agents, such as dithiothreitol, they found that methidiumpropyl-EDTA · Fe(II) cleaved DNA in a non-sequence specific manner. Subsequently, the sequence specific distamycin-EDTA · Fe(II) complex was prepared[149)]. The use of EDTA bound to distamycin, a tripeptide, facilitated the cleavage of DNA in adenine and thymine rich regions. Distamycin can be coupled to either the amino terminus or the carboxy terminus of the tripeptide. These two forms of EDTA cleave DNA at the same sites but from opposite ends of the DNA strand. In yet a further extension of the work, EDTA was bound to each end of the tripeptide[151)]. Another logical step in this work was to join distamycin through the amino groups with fumaric acid as the coupling unit[156)]. With EDTA at the carboxy end of the bis-distamycin it became possible to cleave DNA into 8 to 9 contiguous base pairs of A-T rich double helical DNA. In addition to being used to cleave DNA, methidiumpropyl · EDTA · Fe(II) has been used to cleave ribosomal RNA[154)]. Additional methods of directing EDTA · Fe(II) to DNA have involved coupling the chelate to oligonucleotides[153, 155)]. By this procedure it has been possible to achieve cleavage of complementary nucleotide sequences.

By using various nucleic acid-sequence-directed EDTA · Fe(II) chelates it will become possible to advance genetic manipulation technology and possibly develop novel chemotherapeutic agents.

### *3.1.3 Iminodiacetic Acids*

Iminodiacetic acids have a long record of study in medicinal chemistry, starting over 25 years ago when radiolabelled forms were shown to be taken up into newly formed bone[157,158]. The uptake into bone was attributed to chelation whith calcium and was at a maximum when the calcium chelates had a formation constant of around log $\beta = 5$.

In more recent years a wide variety of iminodiacetic acids, perhaps in excess of 65, have been used to chelate predominantly $^{99m}$Tc[159–163], although $^{103}$Ru has also been screened[164]. The structures of these chelating agents have been selected to achieve a biodistribution of the chelated radiocation to a variety of organs. In some cases studies with $^{14}$C-labelled iminodiacetic acids have revealed clearance patterns which are not displayed by the $^{99m}$Tc chelates: a predominantly urinary clearance was noted for $^{14}$C–N-(2,6-dimethylphenylcarbamoylmethyl)iminodiacetic acid, whereas the $^{99m}$Tc chelate underwent predominantly hepatobiliary clearance[160]. As expected $^{99m}$Tc-N-methyl-iminodiacetate exhibited a rapid renal clearance[159].

In an extensive study by Chiotellis and Varvarigou[162] the position of substituents on the benzene ring and lipophilicity were evaluated for a series of $^{99m}$Tc-iminodiacetates. Examination of Table 3 shows that clearance to the intestine is a characteristic of the more lipophilic molecules. In a few cases, $R_1 = R_2 = R_4 = R_5 = H$, $R_3 = 0 \cdot$ allyl or iPr, appreciable clearance to urine also occurs along with high clearance to the intestines. For the chelates $R_1 = R_2 = R_4 = R_5 = H$, $R_3 = O$ or $O \cdot nBu$ the clearance to the

**Table 3.** Percentage Distribution of $^{99m}$Tc-labelled acetanilidoiminodiacetates in mice 30 min after administration

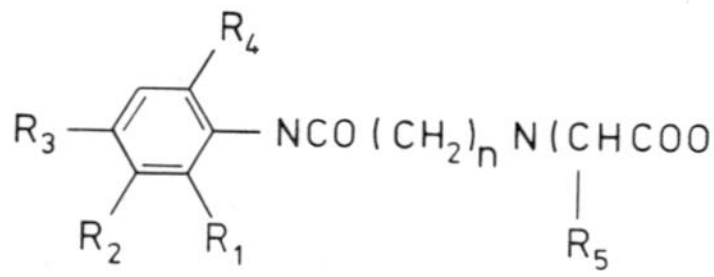


| $R_1$ | $R_2$ | $R_3$ | $R_4$ | $R_5$ | n | Urine | Blood | Liver | Intestines |
|---|---|---|---|---|---|---|---|---|---|
| H | H | H | H | H | 1 | 45 | 1 | 5 | 30 |
| COOH | H | H | H | H | 1 | 28 | 5 | 7 | 35 |
| H | OH | COOH | H | H | 1 | 50 | 5 | 8 | 10 |
| H | H | $NHCOCH_2N(CH_2COOH)_2$ | H | H | 1 | 55 | 2 | 1 | 5 |
| n · Bu | H | H | H | H | 1 | 12 | 4 | 8 | 48 |
| H | H | n · Bu | H | H | 1 | 5 | 4 | 18 | 50 |
| H | H | n · Bu | H | H | 1 | 5 | 1 | 6 | 75 |
| H | H | t · Bu | H | H | 1 | 7 | 2 | 10 | 68 |
| O · nBu | H | H | H | H | 1 | 20 | 2 | 18 | 50 |
| O · nHex | H | H | H | H | 1 | 5 | 5 | 8 | 55 |
| H | H | O · Allyl | H | H | 1 | 25 | 2 | 2 | 50 |
| H | H | i · Pr | H | H | 1 | 30 | 2 | 5 | 45 |
| H | H | O · nBu | H | H | 1 | 5 | 1 | 10 | 60 |
| H | H | O · iBu | H | H | 1 | 5 | 2 | 8 | 75 |
| H | H | O · nHex | H | H | 1 | 1 | 1 | 45 | 50 |
| H | O · nBu | H | H | H | 1 | 2 | 1 | 5 | 78 |
| $CH_3$ | H | H | $CH_3$ | H | 2 | 5 | 1 | 7 | 35 |
| $CH_3$ | H | O · nBu | H | H | 1 | 4 | 1 | 6 | 70 |
| $C_2H_5$ | H | H | $C_2H_5$ | H | 1 | 15 | 4 | 5 | 60 |
| H | H | H | H | $C_6H_5$ | 1 | 20 | 3 | 15 | 48 |
| H | H | H | H | nBu | 1 | 11 | 3 | 12 | 43 |
| $CH_3$ | H | H | $CH_3$ | nBu | 1 | 12 | 3 | 28 | 38 |

intestines is rather similar as is the retention in the liver; yet only slight increase in lipophilicity, $R_3$ = O · nHex, is marked by a high retention in the liver. In general ortho-substitution favoured rapid urinary clearance rate.

In an extension of their study, Chiotellis and Varvarigou[162)] also screened a variety of other iminodiacetates ranging from N-benzhydryl carbamoyliminodiacetate to N-decyl-iminodiacetate (Table 4).

**Table 4.** Distribution, % dose per samples, of N-substituted iminodiacetates of the general formula $R{-}NHCOCH_2N(CH_2COOH)_2$

| Chelate | Urine | Blood | Liver | Intestines |
|---|---|---|---|---|
| Benzhydryl | 8 | 2 | 20 | 48 |
| Phenothiazine | 20 | 5 | 18 | 35 |
| Iminostilbene | 21 | 5 | 18 | 25 |
| Hexyl | 6 | 2 | 15 | 45 |
| Octyl | 6 | 5 | 20 | 50 |
| Cycloctyl | 10 | 1 | 15 | 50 |
| Decyl | 9 | 2 | 35 | 48 |
| N-Decyliminodiacetate | 18 | 5 | 40 | 20 |

Graham et al.[163)] have evaluated a series of 1,5-substituted N-(3-cyano-4-methyl-2-pyrrylcarbamoylmethyl)-iminodiacetates. Examination of their results shows that $N^1$-methylation increases the retention of $^{99m}Tc$ in the carcass and decreases the clearance through the liver to the intestines. At first consideration this might be expected to increase lipophilicity and thus increase hepatobiliary clearance. The authors proposed that intramolecular hydrogen bonding between the N-1 hydrogen and the carbonyl at C-2 occurred and that this blocked interaction of the carbonyl group with water. In the N-methyl series it was proposed that reduced protein binding occurred and that this favoured reduced biliary excretion. It is possible that the increased basicity of the N-methyl series favoured protein association. As little protein is cleared through the bile duct it can be assumed that chelates in the N-methyl series will not be cleared through the bile duct. For both the substituted and unsubstituted N-1 series an increase in molecular weight of the aliphatic moiety C-5 enhanced biliary excretion.

### *3.1.4 Polyaminopolyphosphonic Acids*

Polyaminopolyphosphonic acids have attracted intermittent interest since the early applications of chelating agents in medicine. First synthesized 30 years ago, ethylenediaminetetraphosphonic acid (EDTMP) has only recently been examined for ability to inhibit dental calculus and caries in dogs and rats.

The observations that pyrophosphate and the geminal 1,1-bisphosphonate chelates of $^{99m}Tc$ are taken up by bone led to the polyaminopolyphosphonates also being screened for bone scintigraphic potential. Subramanian et al.[165)] reported that EDTMP chelates of $^{111}In$ and $^{113m}In$ had bone uptake comparable to that of $^{85}Sr$. EDTMP was shown to be marginally superior to diethylenetriaminepentamethylene phosphonic acid (**23**, R = $CH_2PO_3H_2$, DTPMP). The nitrilotrimethylenephosphonic acid chelates were poor skeletal scintigraphic agents in the rabbit.

DTPMP is no more effective than DTPA at removing skeletally incorporated $^{239}Pu$[166]. Even less effective is a partially lipophilic form of polyaminopolyphosphonic acid (**23**, R = hexyl)[167].

Taliaferro and Martell[168] have presented a thorough study of the stability constants of N,N′-bis(2-hydroxybenzyl)ethylenediamine-N,N′-bis(methylenephosphonic acid mono-alkyl ester) (**24**) with copper(II), nickel(II), cobalt(II), zinc(II) and iron(III). Of particular note are the Fe(III) chelates of the ethyl and methyl esters which are soluble and 100% formed at physiological pH. These solubility characteristics commend them for further study in the therapy of iron overload. In contrast, the Fe(III) chelates of the non-esterified ligands are insoluble. Esterification with an oligoethyleneglycolmonoethyl ether could increase hepatobiliary clearance and facilitate imaging of the liver and bile duct with $^{99m}Tc$, $^{67}Ga$, $^{111}In$ or $^{113}In$. Magnetic resonance imaging using chelated gadolinium as a contrast agent might also be possible.

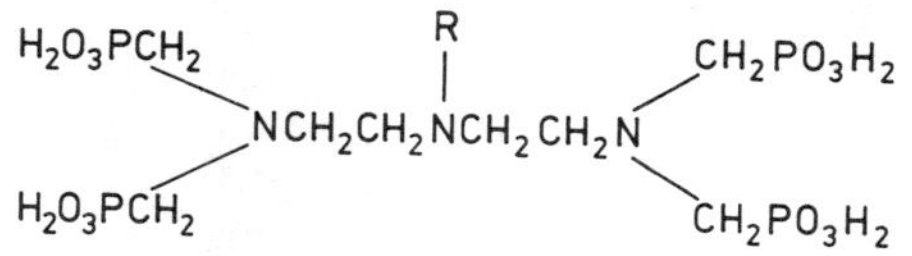


**23**

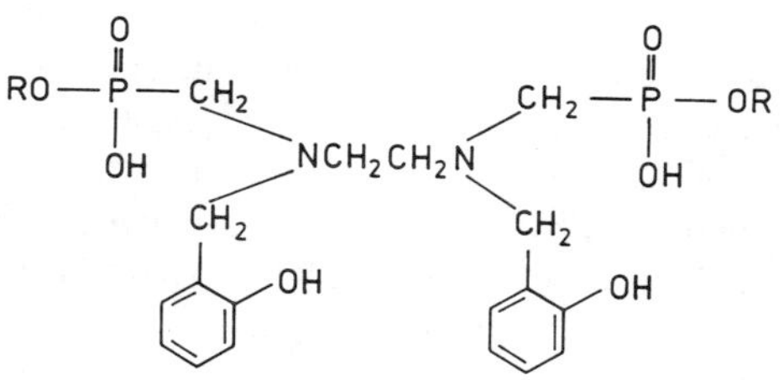


**24**

### *3.1.5 1,1-Bisphosphonates*

The 1,1-bisphosphonates ($>C(PO_3H_2)_2$) have attracted considerable interest in the last 15 years or so because of their uptake by bone, an uptake which is a consequence of their structural similarity to bone-seeking pyrophosphate – see Russell and Fleisch[169] for a review.

The early promise of pyrophosphate as a modulator of the biochemistry of bone was not fulfilled because of its ready hydrolysis. As molecular models indicated that 1,1-bisphosphonates might exhibit some pharmacological responses in bone, a programme of syntheses commenced and this still continues (Table 5).

A detailed account of the biochemistry of the bisphosphonates has recently appeared elsewhere[170]. Briefly, the bis-phosphonates inhibit soft tissue calcification. Bis-phosphonates, such as HEBP inhibit the mineralization of cartilage, bone and dentine. Prolonged administration of HEBP to man at oral doses of 10 $mg^{-1}$ $kg^{-1}$ for more than one month affects the mineralization of hard tissues. Slight changes in the substituents on the gemi-

**Table 5.** Structures of some representative bisphosphonates

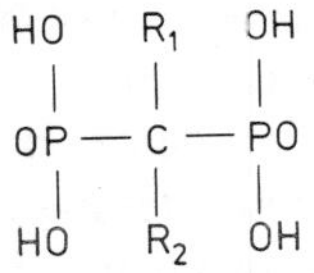

| 1,1-Bisphosphonate | $R_1$ | $R_2$ |
|---|---|---|
| Methylene[a] | H | H |
| Hydroxymethylene[b] | OH | H |
| Ethylidene | H | $CH_3$ |
| 1-Hydroxyethylidene[c] | OH | $CH_3$ |
| 1-Hydroxypropylidene | OH | $CH_2CH_3$ |
| 1-Hydroxybutylidene | H | $(CH_2)_2CH_3$ |
| Dichloromethylene[d] | Cl | Cl |
| Dibromomethylene | Br | Br |
| Difluoromethylene | F | F |
| Carbonyl[e] | – | =O |
| 2,3-Dicarboxypropyl[f] | $CH(COOH)CH_2COOH$ | H |
| 3-Amino-1-hydroxypropylidene[g] | OH | $(CH_2)_2NH_2$ |
| 6-Aminohexylidene[h] | H | $(CH_2)_6NH_2$ |
| 1-Amino-1-phenylmethylene | $C_6H_5$ | $NH_2$ |
| 3-Aminopropylidene | H | $(CH_2)NH_2$ |
| N-Methylaminomethylene[i] | $NCH_3$ | H |
| N,N-Dimethylaminomethylene[j] | $N(CH_3)_2$ | H |
| 1-Aminoethylidene | $NH_2$ | $CH_3$ |
| t · Butylhydroxymethylene | $(CH_3)_3C$ | OH |
| α-Hydroxybenzylidene | $C_6H_5$ | OH |

[a] MBP; [b] HMPB; [c] HEBP; [d] $Cl_2$MBP; [e] CBP; [f] DCBP; [g] AHPrBP; [h] AHexBP; [i] NMBP; [j] DMBP

nal carbon results in varied responses. Unlike HEBP, $Cl_2$MBP treatment does not cause fractures. In spite of their close basic similarities to HEBP, aminobisalkanephosphonates also do not induce fractures. In addition, they are more effective at inhibiting bone resorption. It has been suggested that the differences in physiological properties of the bisphosphonates arise from the ability to form bidentate-bidentate and bidentate-tridentate bridged structures[171)]. No doubt the affinity of bisphosphonates for calcium must be considered an important factor in their biochemical actions.

Bisphosphonates have several important clinical applications which can be divided roughly into two categories. Firstly, their use in regulation of calcium kinetics. Secondly, the exploitation of their bone-seeking properties to carry $^{99m}$Tc and other γ-emitting radionuclides to bone. In the first category, the bisphosphonates are now extensively used to control diseases with increased bone destruction. Foremost amongst these diseases is Paget's disease, a condition of uncertain aetiology and which is marked by excruciating pain in the bones. In many patients treated with HEBP at the rate of 20 mg $kg^{-1}$ $d^{-1}$ normal values of alkaline phosphatase and hydroxyproline can be achieved within 3 to 6 months. This response can last for up to 2 years[172)]. Recently 6-AHexBP as its ester, sodium salt and copper chelate has been administered at 5–20 mg $d^{-1}$, by intravenous injection to suppress osteolysis in patients with Paget's disease and metastatic mammary carcinoma[173)].

Preliminary clinical findings established that this treatment was well tolerated and did not affect the levels of calcium and phosphate in blood. Promising results have been reported for the use of HEBP in primary hyperparathyroidism and tumoral osteolysis, two diseases also characterized by increased bone resorption. In the treatment of ectopic calcification the use of bisphosphonates has not fulfilled the early hopes which were held for them. However, the bisphosphonates could prove useful in the prevention of heterotropic calcification as found in patients with spinal cord injury or in the treatment following total hip replacement. Good results have also been obtained against dental calculus and this had led to the use of bisphosphonates in toothpastes. The anti-arrhitic properties attributed to bisphosphonates have prompted an expansion of the syntheses of bisphosphonates. 4-Chlorophenylthiomethylene bisphosphonic acid has undergone some preliminary screening for anti-arthritic activity[174)].

It is in the use of radiopharmaceuticals that the bisphosphonates have led to an extensive investigation of structure-activity relationships. If additional bonding through the R substituents on C-1 is possible the structure changes from a bidentate-bidentate bridged structure to a bidentate-tridentate bridge[175)]. Bisphosphonates with –OH as a substituent at C-1 have an enhanced affinity for calcium whereas nitrogen-containing ligands have a lower affinity for calcium. Deutsch and Barnett[176)] and Francis et al.[177)] have proposed that the tridentate binding through the hydroxyl and $^{99m}$Tc-HMBP complexes is an important factor in the skeletal uptake of these imaging agents.

Shyre et al.[178)] have suggested that the chelating groups of $^{99m}$Tc-bisphosphonates must be at a distance of 30 nm apart in order to associate with the calcium ions on the hydroxyapatite surface. The best visualization of the skeleton will be achieved by a complex which has a high uptake into new bone with a low uptake into soft tissues. Diseased states of the bone are characterized by the excessive uptake of $^{99m}$Tc into newly formed bone. Subramanian et al.[179)] have examined several $^{99m}$Tc-bisphosphonates for percentage uptake into whole skeleton, the normal bone-to-muscle ratio and the new bone-to-normal bone ratio. An examination of Table 6 shows that while a high uptake into skeleton gives a high normal bone-to-muscle ratio it does not produce a good image of new bone.

**Table 6.** Location of a series of $^{99m}$Tc-bisphosphonates between skeleton and muscle

| $^{99m}$Tc-chelate | % chelate in whole skeleton | Normal bone / Muscle | New bone / Normal bone |
|---|---|---|---|
| HMBP | 58 | 435 | 3.43 |
| DCBP | 52 | 260 | 3.27 |
| NMBP | 31 | 488 | 4.35 |
| AHPrBP | 31 | 407 | 3.79 |
| MBP | 30 | 395 | 4.48 |
| DMBP | 19 | 265 | 6.56 |

The influence of the substituent on the methylenic carbon and uptake into bone has been examined for the bisphosphonate series: MBP, $Cl_2$MBP, CBP, HMBP and DHMBP. As a result of their study, Wang et al.[180)] recommend HMBP for maximum uptake into bone and for fastest clearance from blood.

The influence of the reductant upon the distribution of the $^{99m}$Tc chelate, arising from pertechnetate, has been investigated. Generally, stannous chloride is the preferred reductant although others such as sodium borohydride, formamidinesulphinic acid and antimony(III) have been used. Recently, some interest has also been shown in electrolytic reduction. High performance liquid chromatographic separation on an anion-exchange column has resolved $^{99m}$Tc-HEBP, produced by reduction with sodium borohydride, into 13 complexes with charges ranging from $-15$ to $-8$[181]. Five of the isolated fractions were injected into rats and the uptake into new bone monitored. Small, low-charged complexes – presumably monomeric forms – had a high uptake into bone, whereas the highly charged and presumably polynuclear forms had low uptake into new bone[182]. It has been proposed that $^{99m}$Tc-bisphosphonates with a low protein binding become more localized in bones than the complexes which have a stronger protein association[183]. $^{99m}$Tc-bisphosphonates bind to plasma proteins less avidly than the pyrophosphates and become more localized in bone than pyrophosphate forms[184].

The role of stannous ions in the $^{99m}$Tc chelate have been examined and it has been shown that they are not essential for bone uptake[185].

Palliative treatment of osseous neoplasms is an important component of the treatment schedule for cancer. Radiophosphorus-containing HEBP has been examined for this role. Clinical evaluations of $^{32}$P-HEBP have established that the early promise of its efficacy was unfulfilled[186]. An alternative radionuclide is $^{33}$P, a less energetic isotope. The similarities in the chemistry of technetium and rhenium have led to consideration of $^{186}$Re-HEBP as an alternative palliative theory[187]. Laboratory studies with the rat have confirmed the potential of this treatment. As $^{117m}$Sn, $t_{1/2}$ 14.03 d, has an abundance of short range conversion and Auger electrons a radiotherapy for bone tumours based upon $^{117m}$Sn-HEBP has been proposed[188].

### 3.1.6 8-Hydroxyquinolines

Interest in the oxines has for long been intense and in the early years many new oxines were synthesized. Recently Roth and Höfkens[189] have reported the synthesis of a series of N,N,N′,N′-tetrakis-(8-hydroxy-5-quinolymethyl)-α, ω-diaminoalkanes. Examples of their use in the medical sciences are still awaited.

The interest in oxines has now moved from antimicrobial therapy to nuclear medicine and the chemotherapy of malaria. A few other applications have also been found: 5,7-diiodo-oxinate enhances gastrointestinal uptake of zinc, an element whose deficiency may cause acrodermatis enteropathica[190]; several halogenated oxines – 5-chloro-7-iodooxine (Entero-Vioform), 5,7-diiodooxine (Diodoquin) – have been used in the treatment of amoebic dysentry but this is now a discontinued therapy. 5-Methyloxine has proved an effective therapy in the treatment of cholera in humans[191, 192].

The deduction that oxygen does not mediate to any significant degree in the ATP-generating electron system in the malaria parasite but instead participates in biosynthetic reactions through metalloproteins oxidases is a significant departure from the metabolism of mammalian host cells. This observation has led to an examination of the interaction of lipophilic chelators with the malaria parasite. To date a number of chelating agents – hydroxamic acids, catechols, dithiocarbamic acids and substituted 8-hydroxyquinolines have been screened as anti malarials[193–195].

Scheibel and Adler[195] have shown that the anti-malarial effects on the growth of the parasite appear to correlate with the ability of 5,7-dichlorooxine, 5-chloro-7-iodooxine, 5-chlorooxine, 6-hydroxy-1,7-phenanthroline, and 5-chloro-7-(3-diethylaminopropylaminomethyl)-8-quinoline didydrochloride to chelate metals.

The addition of an electron-withdrawing nitro or chloro groups at position C-5 of oxine might be expected to increase the ionization of the phenolic group and thus improve chelation. However, substitution at positions C-5 and C-7 do not imbue the oxines with a plasmodiacidal activity which is superior to unsubstituted oxine. Enhancement of lipid solubility of the oxine series alone does not confer enhanced plasmodiacidal activity as determined for 6-hydroxy-1,7-phenanthroline, a finding which is contrary to antibacterial effects which parallel partition coefficients for organic-aqueous systems[196, 197]. However, the incorporation of a halogen at C-5 and a lipophilic substituent at C-7, as in 5-chloro-7-(3-diethylaminopropylaminomethyl)-8-quinolinol dihydrochloride, combines both enhanced chelation and enhanced lipophilicity and this is marked by plasmodiacidal activity which exceeds that of the other oxines studies[193]. 5-Chloro-7-(3-diethylaminopropylaminomethyl)-8-quinolinol dihydrochloride has long been recognized as an effective luminal amoebicide which is well tolerated by animals.

*In vitro* plasmodiacidal activity has also been noted for 2-mercaptoquinoline-N-oxide and for 5-methyl oxine which has been used in humans to control cholera[191, 192]. The antimalarial activity of the oxines is, unlike their antibacterial activity, insensitive to cobaltous antagonization[195]. This inability of cobalt to suppress antimalarial activity points to an inhibitory mechanism which does not depend upon peroxide formation.

The well-established chelating properties of oxine have been exploited in NIDM. Oxine is an effective agent for transporting $^{111}In$ into leucocytes and granulocytes which are then used to detect abscesses[198]. Other developments are the radiolabelling of platelets with the oxine of $^{68}Ga$, for detecting thrombosis or artherosclerosis[199]; the radiolabelling of lymphocytes with $^{111}In$ for detection of cardiac rejection[200]; lymphography with $^{99m}Tc$-Sn oxine radiocolloids[201].

## 3.2 *Macromolecular Chelating Agents*

Investigations of the complexation of polyvalent cations by biomacromolecules, particularly metalloproteins, have obviously been conducted for many years. Investigations of the binding of polyvalent cations by synthetic polymers are more recent. The chemistry of macromolecular chelating agents (MCAs) has been reviewed by Hodgkin[202], Kaneko and Tsuchida[203] and Sahni and Reedijk[204].

The MCAs are represented by a wide variety of structures:

(i) Mercaptodextrans, disulphide-reducing polythiols formed by thiolation of dextran[205];

(ii) various polyacryls bearing DFOA[206]; polyvinylamine-vinyl (2,3-dihydroxy-5,6-dimethylbenzenesulphonamide)-vinyl sulphonic acid[207];

(iii) various polysaccharides bearing DTPA[208, 209];

(iv) proteins bearing DTPA or pAPhEDTA (**10**, $n = 0$, $R = NH_2$) or DFOA[28, 210–215].

(v) poly-(EDTA-$cys_2$), formed by polymerizing EDTA-$cys_2$ by production of disulphide bonds[208].

(vi) hydroxamic acids linked by oligomethacroyl units[216–218].

In the succeeding paragraphs the applications of these MCAs are discussed in some detail.

The immobilization of low molecular weight chelating agents on macromolecules received an initial impetus with the synthesis of pAPhEDTA[214)]. Over the years pAPhEDTA, and related forms, have been coupled to a series of macromolecules and other chemicals and evaluated for potential as diagnostic agents for detection of tumours and for studying dysfunction of organs. Diazotized pPhEDTA coupled to proteins such as human serum albumin (HSA) and fibrogen can be monitored in experimental animals and humans through the chelation of $^{111}$In[214)]. The plasma disappearance rate in humans of the HSA-chelate is similar to that found for radioiodinated HSA. Although not a macromolecule bleomycin, a fermentation product with a molecular weight of 1390 Daltons and accumulated by some cancer cells, has been derivatized by reaction with pAPhEDTA to give BLEDTA[219, 220)]. In clinical studies with $^{111}$In-BLEDTA, positive scans were found in 18% of the cancer cases examined[219)]. In 22% of cases, tumour sites were detected where other procedures had failed. Unfortunately only tumours 1.5 to 2.0 cm in diameter were detected.

The pioneering work of Meares and his colleagues led to other procedures for immobilizing chelating agents on macromolecules. Whereas Krescarek and Tucker[117)] used the mixed acid anhydride procedure to couple DTPA to proteins, Bulman and Griffin[208)] reacted DTPA bisanhydride with cyanogen bromide-activated cellulose bearing 1,6-diaminohexane as a spacer-arm. Hnatowich et al.[210)] have now demonstrated that antibodies derivatized with DTPA-bisanhydride retain their capacity for interaction with the antigen. In addition to immobilization of DTPA, the immobilization of DFOA on human immunoglobulin has been achieved by coupling reactions using glutaraldehyde or carbodiimide[212)]. The labelling of monoclonal antibodies with radionuclides such as $^{67}$Ga, $^{111}$In and $^{113m}$In through chelation by immobilized chelating agents will facilitate the detection of specific tumours. In an extension of this procedure it has been proposed that tumours could be destroyed by a radiotherapeutic procedure which carries the β-emitting radionuclide $^{90}$Y into the tumour bound as the DTPA chelate coupled to a monoclonal antibody[221)]. Recently Vaughan et al.[221a)] have issued a cautionary note on the potential of tumour killing by using radiolabelled antibodies. If a 60 Gy dose in 7 days was sufficient for tumour sterilization, this would result in a whole body dose of 14.3 Gy for chelated $^{90}$Y. The maximum reasonable whole body dose is 2 Gy.

The use of MCAs is by no means restricted to radiopharmacy. Brady et al.[222)] have exploited this technique by using as a contrast agent $Mn^{2+}$ bound by DTPA immobilized on the monoclonal antibody antimyosin. The extent of infarcted myocardium was determined with 2 mg of antibody bearing 50 μg $Mn^{2+}$. It is to be expected that many more similar contrast agents await development. Their development will extend magnetic resonance imaging.

In a continuation of studies with pAPhEDTA, it has been shown that monoclonal antibodies raised against antigens bearing a chelated polyvalent cation exhibit a high specificity for the trivalent metal ion. Replacement of indium(III), for instance, by scandium(III) or gallium(III) can decrease the antibody-binding constant by more than three orders of magnitude[223)].

The uses of MCAs in medicine, other than in NIDM, are rather limited. MCAs have been examined for ability to suppress gastrointestinal (GI) uptake of radiocations and cations of toxic elements. In the 1960s alginates, incorporated into bread, were shown to suppress GI-uptake of $^{90}Sr$ in experimental animals[224]. Several years later it was demonstrated that DTPA immobilized on cellulose also brought about suppression of GI-uptake of alkaline earth radiocations[225].

Although not in the strictest sense MCAs, the success of thiol-bearing polymers used by Bakir et al.[226] to suppress enterohepatic recirculation of methylmercury should be noted. The patients in this study were Iraqi farmers and other villagers who consumed seed grain which had been treated with a mercury-based fungicide.

A general lack of effectiveness has been noted for several MCAs in the suppression of GI-uptake of cadmium in mice. In this study the MCAs, the metal-binding capacity was 2–6.8 mM $g^{-1}$, ranged from DTPA immobilized via amide bond formation to finely divided aminopropyl-silica, xanthates of polyvinylalcohol and various polysaccharides, and a dextran bearing ethylthiol moieties produced by action of ethylenesulphide on dextran[209].

Three basic types of MCAs have been prepared in a search for countermeasures to iron-overload. In 1976 Ramirez and Andrade[206], motivated by the high incidence of accidental iron-poisoning in children, prepared polymer-DFOA grafts for use in extracorporeal detoxification of acute iron poisoning. In the search for therapeutic countermeasures to iron overload, Dawson et al.[207] prepared an iron-binding MCA by reacting 2,3-diacetoxy-5,6-dimethylbenzenesulphonyl chloride with poly(vinylamine-vinylsulphonate). In passing, it should be noted that Durbin et al.[227] have examined oligomers bearing pendant catechols for ability to enhance the clearance of $^{239}Pu(IV)$ from mice. These polymers were less effective than the monomeric forms.

As an alternative to polymers bearing pendant catechols, Winston and his colleagues have prepared hydroxamic acid polymers in which hydroxamate side-chains were linked by oligomethacroyl units[216–218]. From experiments on iron-overloaded mice four polymers were found to be as good as, or better, than DFOA in removing iron. Several of the polymers had quite low toxicity. A polymeric form of **(25)** of around $10^6$ Daltons was particularly effective.

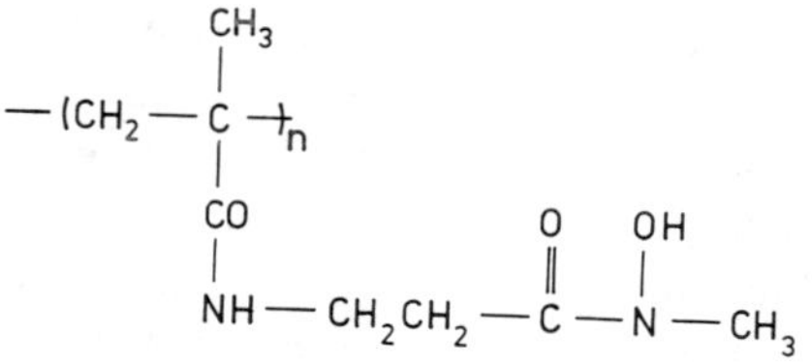


**25**

Immobilized MCAs have applications in several fringe areas of medical sciences:

(i) they have been used to remove aluminium ions from ATP and thus correct some anomalous kinetic data[228];

(ii) when linked with inductively coupled plasma spectroscopy they have aided in speciation studies of arsenic, chromium, selenium and tellurium[229];

(iii) they have been used in purification of metalloproteins by metal-affinity chromatography[230, 231];

(iv) they have formed the basis of an alternative to radioimmunoassay by using fluoroimmunoassay of immobilized terbium and europium[232–235].

The commercial potential of fluoroimmunoassay is extensive, as judged by the investment made by LKB who now sell the equipment.

## 3.3 Chelating Agents of Microbial Origin and Their Biomimetic Counterparts

For many years now it has been established that microorganisms are important sources of many chelating agents, particularly ferric-selective ones[236]. Chelating agents of microbial origin range from what could be considered as a substituted iminodiacetic, that is pyridine-2,6-dicarboxylic acid, through the calcium-specific ionophores to the ferric-selective ones, chemicals which generally possess either catechol or hydroxamic acid moieties. The siderophores have served as the basis for the extensive biomimetic syntheses programme conducted by Raymond and his colleagues, covered in several reviews[237–239], and also by others[240–242].

### 3.3.1 Ionophores Specific for Divalent Cations

Predominant in this category are the calcium-specific ionophores calcimycin (A23187), lasalocid (X537A) and ionomycin, one of the few naturally-occurring 1,3-diketones. These ionophores have been used in investigations of the role of calcium in secretory cells such as mast cells and β-pancreatic cells – see Campbell[110] for an account of the regulation of the inorganic biochemistry of calcium by ionophores. Although it complexes several divalent cations, antibiotic M144255, isolated by Brooks et al.[243], has a high affinity for zinc and thus has been given the trivial name zincophorin. Zincophorin, like several other ionophores, exhibits anti-coccidial activity against *Eimcria tenella* in chicks[243]. It is also active against Gram-positive bacteria.

### 3.3.2 Siderophores

The hydroxamates have been the more extensively investigated of the two kinds of siderophores. For many years now DFOA, **26**, has been used in the treatment of siderosis such as the transfusional form induced while treating β-thalassaemics.

In spite of its almost prohibitive cost – £ 3.12 per gram in 1978[21] and that it can not be taken orally – DFOA is playing an important role in health care. Under good clinical management DFOA is able to bring the young thalassaemic into negative iron balance[244]. If this treatment schedule was maintained, medication could cost around £ 2,500 per annum for a lifetime[21]. Currently research into synthetic ferric-selective chelating agents has become extensive. The ineffectiveness of DFOA in removing aged deposits of iron has lead to investigations into new methods of drug delivery as well as modifications of the structure. Liposomally entrapped DFOA was an ineffective therapy when administered orally or via the intraperitonal cavity, but when given intravenously the treatment doubled the clearance of $^{59}$Fe from mice[245]. In clinical trials of DFOA entrapped in red cell ghosts Green et al.[246] achieved a much greater clearance of iron than that achieved by slow subcutaneous infusion of DFOA. This delivery system was particularly effective in patients who were the least iron-overloaded.

Modifications of the structure of DFOA have included the synthesis of N-stearoyl-DFOA[247], a lipophilic form which undergoes clearance through the bile duct of the perfused liver, and also catechol derivatives (**26**, R = 2,3-dihydroxylbenzoyl and 2,3-

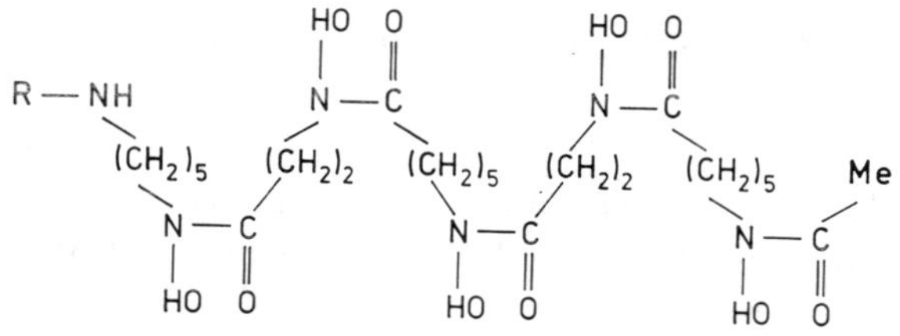

**26**

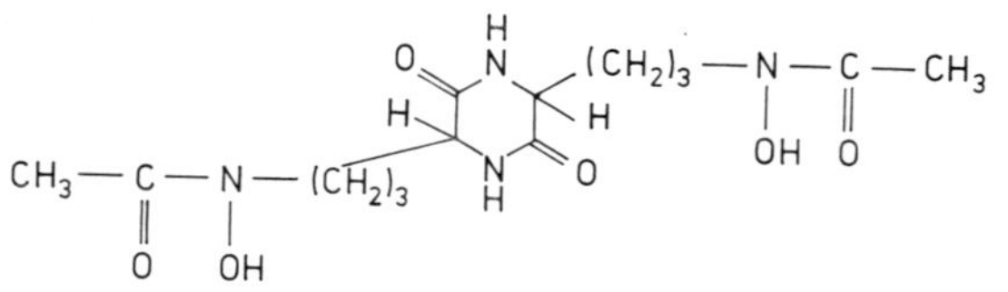

**27**

dihydroxy-4-carboxyl-benzoyl)[248]. Both of these catechol derivatives were more superior than DFOA at removing iron from ferric transferrin and were only slightly less effective than tricatechol ligands discussed below. Interestingly the dimethyl protected precursor of 2,3-dihydroxy-4-carboxybenzoyl-DFOA removed iron from transferrin at the same low rate as DFOA and thus confirmed the role of the 2,3-dihydroxy moiety in overcoming the kinetic barrier of DFOA in removing iron from transferrin.

In addition to being used to mobilize iron, DFOA has been examined as an anti-inflammatory agent for the treatment of rheumatoid arthritis which may be precipitated by the increased stores of iron which accompany hyposideraemic anaemia[77]. The mobilization of iron from patients with rheumatoid arthritis indicates that it may be possible to limit damage to the synovial membrane, by oxygen free radicals of oxygen, by removing iron deposits from the synovial membrane.

Soon after the introduction of DFOA to pre-clinical studies it was evaluated for ability to mobilize $^{239}Pu(IV)$ in experimental animals. As it was not superior to DTPA there were no further studies with it. In a lipophilic form N-stearoyl DFOA was also ineffective as a mobilizing agent for $^{239}Pu(IV)$, as was rhodotorulic acid **(27)**[121]. The greater chemical stability of rhodotorulic acid and the ease with which it can be obtained in high purity has resulted in it being evaluated as a potential agent for treating transfusional siderosis. Grady et al.[249] have shown rhodotorulic acid to be only 16% more effective than DFOA in humans. As it causes local inflammatory reactions it is unlikely to replace DFOA.

The extensive reliance upon renal dialysis to prolong the lives of people suffering from kidney failure has led to diseases caused by uptake of aluminium from the dialyzate. The use of aluminium-free dialyzate has led to many fewer reports of dialysis dementia and dialysis osteomalacia. Several studies have demonstrated the clinical efficacy of DFOA in combatting these two diseases[250–252]. Investigation of bone samples removed from the iliac crest before and after treatment for osteomalacia have given conflicting results on the mobilization from aluminium in bone. By energy-dispersive X-ray microanalysis Brown et al.[253] showed no change in aluminium content of the bone biopsies. In contrast, Day et al.[254] found significant decreases in the aluminium content of both trabecular bone and marrow in five out of five patients. In this treatment the patients received thrice-weekly infusions of 6 g of DFOA into a blood line of the dialyzer. Treatment for 6–10 months resulted in the clearance of 300–1500 mg of aluminium.

The response of bacteria to iron-depleted media is well-characterized as the release of the siderophores. Less well known is the response of microorganisms in rich media to the presence of toxic cations. Premuzic et al.[255] have investigated the response of *Pseudomonus aeruginosa* CSU and *P. aeruginosa*-PAO-1 to uranyl and thorium salts added to iron-deficient and iron-supplemented media. Analyses of the ultrafiltered culture media showed that both actinides induced the formation of complexing agents. In the presence of iron, both uranyl and thorium salts induced the formation of chemicals containing hydroxamate and phenol catechol functions. Preliminary analyses of a thorium-induced chemical indicated it was a pyochelin. It is possible that these thorium-binding chelating agents could be used to mobilize intracellularly accumulated plutonium.

As Raymond, the principal advocate of the biomimetic approach to the synthesis of new ferric-binding agents has reviewed this topic, only a brief discussion is justified here[238, 239]. The greater part of this work has concentrated on synthesizing stable analogues of enterobactin (**28,** X = O), a cyclic triester of 2,3-dihydroxybenzoyl-L-serine. Enterobactin is readily hydrolyzed and thus is of no value in chelation therapy. Attempts to synthesize the trilactan analogue of enterobactin (**28,** X = NH) have failed[256].

These new siderophores are identified by acronyms based upon CAM derived from catecholamide which results from substituting 2,3-dihydroxylbenzoyl or its derivatives (**29**) on a series of polyamines. The introduction of a carboxylic moiety at C-4 or a sulpho moiety at C-5 in the benzene ring extends the acronym to CAMC (**29,** R = –COOH, $R_1$ = H) or CAMS (**29,** R = H, $R_1$ = $-SO_3H$). The introduction of these two substituents leads to increased oxidation resistance of the catechol functions as well as increasing the hydrophilicity[257]. The participation of the hydroxy group at C-3 in chelation was established by preparation of an equivalent salicylamide siderophore[227]. Around 30 CAM-type siderophores have now been reported.

As the phenolic groups are weak acids they do not dissociate at pH 7[258] and thus do not form complexes with large weakly-charged cations. With small strongly-charged cations, Fe(III), Pu(IV) and Am(III), deprotonation of the phenolic groups of catechol occurs at pH $<7$. By judicious selection of polyamines, the chelation of Ga(III), Fe(III),

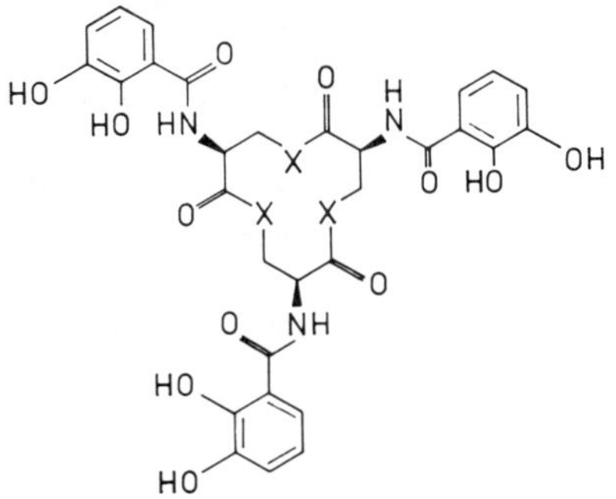


**28**

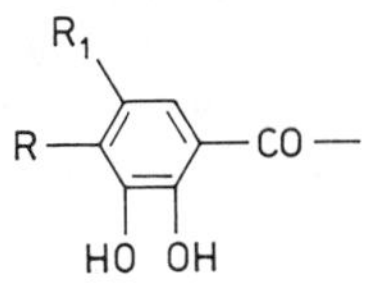


**29**

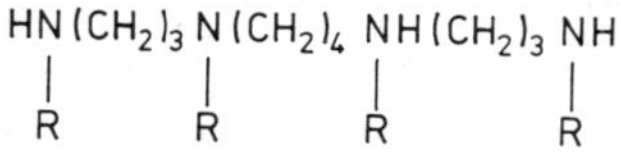

**30**

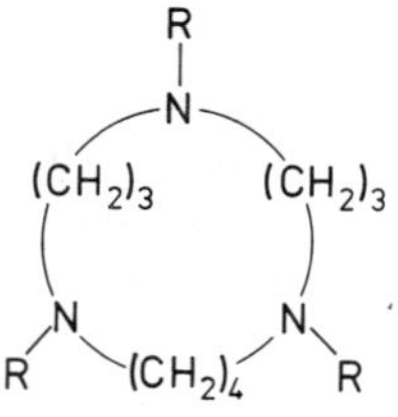

**31**

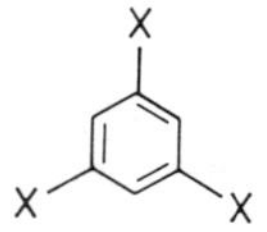

**32**

Pu(IV) and Am(III) at the appropriate denticity can be achieved. The polyamines used have been linear or aromatic polyamines or cyclic polyamines. For the linear polyamines ligand (**29**) is separated by 3 and 4 methylene units or 3, 4 and 3 methylene units, thus resulting in the acronyms 3,4-LICAM[254], 3,4-LICAMC[260] and 3,4-LICAMS[261], (**30,** 3,4,3-LICAMC, R as in **29**). Alternatively various cyclic polyamines were used e.g., (**31,** R as in **29**) which leads to an acronym series CYCAM[257].

Further variations have been produced by substituting (**29**) on 1,3,5-triaminomethylbenzene to give the MECAM series (**32,** X = $CH_2NHR$; R as in **29**)[257, 262]. Yet further variations were introduced where the basic benzene platform was altered (**32,** X = –NH · CO · $CH_2$–R; R = 2,3-dihydroxphenyl) to give TRIMCAM[257]. Substitution of the sulpho moiety on 2,3-dihydroxyphenyl at C-5 led to TRIMCAMS[257].

In the LICAM series lipophilicity, hence ability to cross cell membranes, was introduced by substitution of alkyl groups on the terminal nitrogens[262]. Rastetter et al.[256] have also produced a linear polycatecholamide. However, unlike the LICAM series their siderophore is a chiral analogue, synthesized from L-asparagine, for which a formation constant of $10^{46.5 \pm 1.2}$ has been calculated for its ferric complex. In contrast the formation constant for ferric enterobactinate is $10^{52}$ [263].

Investigations of the kinetics of the reaction of these new siderophores with iron-saturated transferrin showed a rapid formation of a ternary complex with transferrin, followed by a slow step in which the ferric siderophore was released from the apoprotein. Weitl et al.[257] have evaluated the ferric-chelating properties of several of these siderophores and found the following order of effectiveness for removing iron from transferrin: enterobactin > MECAMS > MECAM > LICAMS > DFOA > TRIMCAMS.

In a departure from the biomimetic catecholamide-based siderophores, Raymond's group have turned to derivatives of 1-hydroxy-2(1*H*)-pyridone a structure which can be regarded as a cyclic hydroxamic acid[264]. Unlike hydroxamate siderophores, 1,5-bis[1,2-dihydro-1-hydroxy-2-oxo-pyridin-6-yl)carbonyl]-1,5-diazaheptane (**33**) rapidly removes iron from transferrin.

The similarities in the chemistry of Pu(IV) and Fe(III) were instrumental in the initiation of this extensive programme of synthesis. Additional applications in nuclear

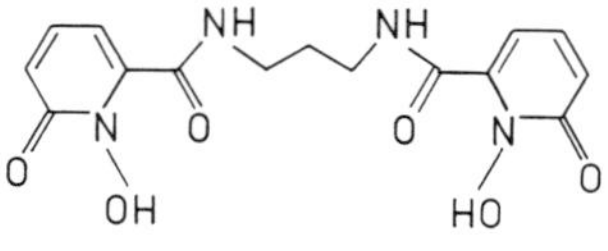

**33**

medicine have also been considered. The chemistry of the biomimetic siderophores used to remove transuranics has been discussed in detail elsewhere[227, 265]. Essentially, the problem is to remove plutonium(IV) and americium(III) from serum, where they are bound to transferrin, and from the parenchymal cells of the liver. An additional target for depletion of the transuranic is the surface of bone where both radionuclides deposit. The decorporation of the skeletally accumulated transuranics is only feasible if chelation therapy quickly follows uptake of the elements. Once the radiocations are incorporated into the bone matrix they become inaccessible to chelating agents until the bone matrix is reworked by the osteoclasts.

The ligands examined for ability to mobilize plutonium and americium varied from the CYCAM, LICAM and MECAM series to a macromolecular form (ca. 3100 Daltons) bearing 14 catecholamide moieties[224]. This extensive screening programme identified $N^1$,$N^5$,$N^{10}$,$N^{14}$-tetrakis(2,3-dihydroxy-4-carboxybenzoyl)-tetraazatetradecane (3,4,3-LICAMC) as the most effective chelator for incorporated $^{239}$Pu. Substitution of alkyl groups on the terminal nitrogens increased lipophilicity but was found to hinder chelation and delay excretion of the radionuclides. In this evaluation of 3,4,3-LICAMC and the other ligands mice and beagles were used[227, 266], whereas Stradling et al.[267] used rats in a comparison of the efficacies of 3,4,3-LICAMC and DTPA for the decorporation of inhaled transportable forms of plutonium and americium. For these inhaled transuranics, the principal forms which could lead to contaminations in the workplace, DTPA proved to be superior to 3,4,3-LICAMC, although both chelating agents were equally effective at clearing the circulating radionuclides from the blood.

The chelation of Pu(IV) and Am(III) by the LICAM series has been studied in detail at neutral pH by electrochemical and spectrophotometric procedures[268]. The Pu(IV) chelate of 3,4,3-LICAMS appears to be a tris (catecholate) complex, indicating that the full denticity of the ligand is not utilized in vivo. Investigation of the complexation of Pu(IV) by 3,4,3-LICAMC establishes a complexation involving the carboxylate para to the carbonyl. Spectroscopic evidence of the complexation of Am(III) by 3,4,3-LICAMS and 3,4,3-LICAMC was also obtained. Significant differences in the spectra of the two complexes were noted. The authors did not exclude complexation through the C-4 carboxylates.

The decorporation of plutonium from mice by 1,5-bis[1,2-dihydro-1-hydroxy-2-oxopyridin-6-yl)carbonyl]-1,5-diazapentane – results cited as "manuscript in preparation"[264] – has been reported.

The biomimetic approach to the synthesis of new siderophores has not been restricted to studies in the USA: Kontoghiorghes[241, 242, 269] has synthesized hydroxypyridones **(34, 35)** and demonstrated their ability to mobilize iron(III) from ferritin. Intragastric administration of 1,2-dimethyl-3-hydroxy-pyrid-4-one **(34)** proved to be as effective as intramuscular DFOA in mobilizing iron from the iron-overloaded rat. This effectiveness of an orally administered chelating agent is particularly noteworthy.

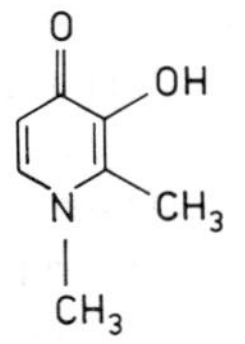

**34**

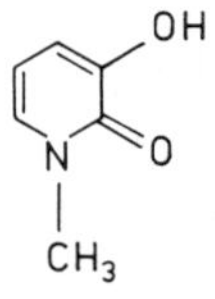

**35**

### 3.3.3 Siderophores in NIDM

The use of DFOA in radiopharmacy has been principally restricted to enhancing the clearance of circulating $^{67}Ga$ which has not been taken up into tumours and abscesses, thus enhancing the images. In human studies DFOA accelerated the clearance of $^{67}Ga$ from the blood but tumour images were not necessarily improved[270]. As its N-succinyl derivative, the DFOA complex of $^{67}Ga$ has been shown to have potential for measuring renal tubular secretion[271].

The potential of the LICAM and MECAM series in nuclear medicine has been demonstrated[272]. The introduction of isopropyl moieties onto nitrogens to give 3,4-DiP-LICAMS and TiP-MECAMS resulted in $^{67}Ga$ and $^{111}In$ chelates which were cleared primarily through the kidneys. In contrast, the less polar 3,4-Di-LICAM chelates were cleared through the liver. As 3,3-LICAMC cleared $^{67}Ga$ from blood, where it is associated with transferrin, it could be of value for reducing the radiation burden from $^{67}Ga$, which is otherwise unsatisfactorily slow to clear[273].

## 3.4 Cyclic Systems

### 3.4.1 Crown Ethers and Cryptands

The synthesis of crown ethers in 1967 fortunately occurred about the same time as naturally occurring lipophilic metal-binding agents were being identified as products of microbial growth. As the literature on synthetic and natural product ionophores is extensive[274, 275], there is no justification for a discussion on their chemistry here.

The search for medicinal applications of crown ethers and cryptands has so far led to only a few pre-clinical studies. Many of the crown ethers and natural ionophores are toxic. Presumably the toxicity arises from the slow leakage of potassium out of cells and of sodium into cells. Antibacterial activity has been noted for several crown ethers and found to parallel potassium transport[276]. The established coccidiostatic activity of ionophores has led to the screening of crown ethers for coccidiostatic activity[277]. As yet the synthetic ionophores have shown no evidence of superiority over their natural product counterparts.

In a search for crown ethers with lowered toxicity levels, several silacrowns (**36**) have been prepared[277, 278]. The introduction of two siloxyl oxygen atoms modifies the binding alkali metal ions, a phenomenon which arises from the lower basicity of the siloxyl oxygen atoms[279]. Investigations of their hydrolytic stability established that they were susceptible to hydrolysis in normal saline. Incorporation of hydrolytically labile bonds into ionophores may lead to decreased but acceptable levels of toxicity and thus ionophores of medicinal value.

In the early 1960s when there was concern over levels of $^{90}Sr$ in the atmosphere various research programmes were directed at producing acceptable decorporation techniques. The problems are obvious: as the divalent cations of calcium and strontium have similar ionic radii – 9.9 nm and 11.3 nm, respectively, the selectivity of chelating agents for strontium must be considerably greater than that for calcium to ensure that calcium is not depleted and that tetany is not induced. All attempts to mobilize radiostrontium from the body without mobilization of calcium have by and large failed. There is perhaps one exception. Müller[280] has examined the macrobicycle 4,7,13,16,21,24-hexaoxa-1,10-diazabicyclo-[8,8,8]-hexacosane (kryptofix 222) which has a selectivity for strontium which exceeds that for calcium. Examination of kryptofix 222 for ability to mobilize strontium from animals has shown that it must be injected within 2–3 min of strontium uptake to achieve its removal.

The chemistry of crown ethers and cryptands has progressed a long way from their status as chemical curiosities. Recently Metcalfe et al.[114] have outlined the development of a fluorinated cryptand as an agent for measuring intracellular [$Na^+$].

The preparation of cryptands (**37**) bearing –$CH_2COOH$ on the ring nitrogens is a predictable development[281]. Investigations of the formation constants formed with the alkaline earths indicates that all nitrogen atoms may be involved in bonding to the cations, whereas perhaps only two nitrogen atoms are involved in the bonding to transition metal ions. As yet these ligands have not been screened for application in the medical sciences.

The extent to which the design of cyclic ligands has now progressed is demonstrated by the synthesis of the endocyclic dihydroxamates, the sexidentate 5,14-dihydroxy-4,15-dioxo-1,5,14,18-tetraaza-8,11,21,24-tetraoxacyclohexacosane (**38,** R = H) and its 1,18-diacetic acid derivative (**38,** R = $CH_2COOH$)[282]. These two endocyclic macrocycles form more stable complexes with nickel(II) and iron(III) than analogous ligands having exocyclic hydroxamate donor ligands. The efficacy of these ligands in reducing iron-overload is awaited with interest.

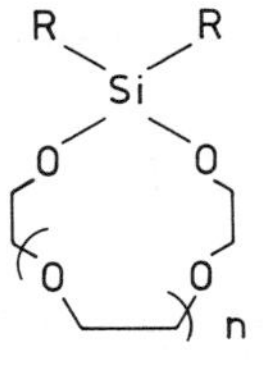

**36**

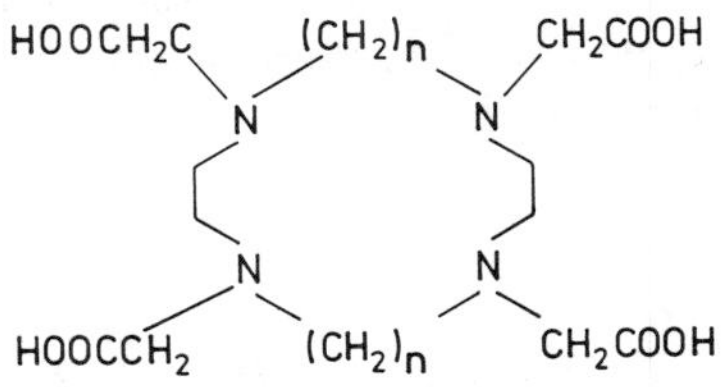

**37**

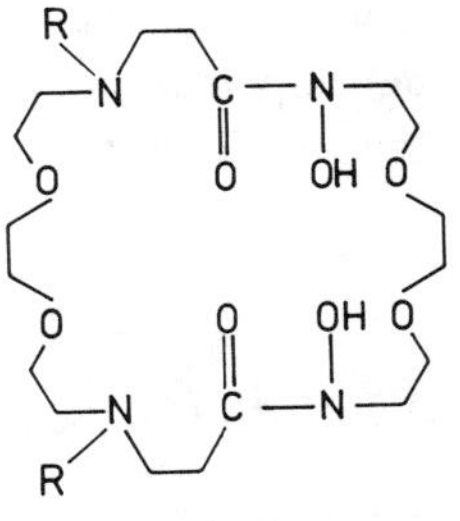

**38**

## 3.4.2 *Porphyrins and Phthalocyanines*

Recent developments in the chemistry of porphyrins and their derivatives have been extensive as various research groups have sought novel oxygen-binding systems for alternatives to blood in surgery. These developments have been adequately covered elsewhere[283].

In addition to the chemistry of porphyrins in relation to oxygen-binding, the chemistry of other metalloporphyrins has developed in recent years. δ-Aminolevulinate (ALA) synthase and haem oxygenase are the rate-limiting enzymes and haem synthesis and degradation respectively. Both are enzymes which are highly inducible and whose activities react to many chemicals. ALA-synthase is of clinical interest because of a genetic disorder, acute intermittent porphyria, AIP, which is associated with neurological impairments. In contrast to marked ALA-synthase in AIP, elevated haem oxygenase activity is of less concern as it may lead to only occasional development of jaundice. In the premature newborn, however, this is a serious condition as it could result in hyperbilirubineaemia-induced neurotoxicity. To-date five synthetic metalloporphyrins – tin(II), cobalt(II), chromium(III), zinc(II) and manganese(II) protoporphyrins – have been examined and proved to be competive inhibitors of haem oxygenase in vitro – see Kappas and Drummond[284]. Sn(II)-protoporphyrin has proved to be the most effective competive inhibitor of haem oxygenase in vitro and also active in suppressing hyperbilirubinaemia in the newborn rat. In limited clinical studies the metalloporphyrin also reduced serum bilirubin levels in experimentally induced jaundice in animals and some forms of jaundice in man.

Although it has long been recognized that intravenously injected porphyrins have an affinity for lymphatic and neoplastic tissues, it is only recently that radiolabelled metalloporphyrins have been screened for lymph node imaging – formerly porphyrins were detected by their fluorescence in target tissue. The studies of Vaum et al.[285] have established that $^{111}$In-labelled tetra(4-N-methylpyridyl)porphyrin has a potential as a lymph node imaging agent.

The similarities of phthalocyanines and porphyrins have led to the screening of $^{99m}$Tc- and $^{67}$Ga-tetrasulphophthalocyanines for ability to detect tumours[286]. Whereas tumour blood and tumour muscle ratios for the Tc-chelate did not reach satisfactory levels within 24 h, satisfactory levels were reached for the Ga-chelate. Although the absolute level of tumour uptake to the Ga-chelate was 5–6 times lower than the absolute uptake of Ga-citrate, rapid blood clearance of the tetrasulphophthalocyaninate permitted an easier identification of tumours. The success of this work will no doubt lead to yet more developments in NIDM. The development of contrast agents based upon porphyrins and tetrasulphophthalocyanines is anticipated.

## *3.5 Chelating Agents of Intermediate Hardness*

### *3.5.1 Polyamines*

The polyamines fall into a category intermediate between hard and soft bases. Their primary use is in the treatment of Wilson's disease. Earlier it has been noted that TRIEN has been used for almost 20 years to treat Wilson's disease[287]. Several other polyamines, both linear and cyclic have also been examined for cupriuretic activity in laboratory animals. Although cyclic polyamines such as 1,4,8,11-tetraazacyclotetradecane (cyclam) have high copper-binding formation constants their enhancement of copper clearance does not merit use in the clinic[288]. Borthwick et al.[289] have proposed 3,7-diazanonane-1,9-diamine (2,3,2-tetraamine, $NH_2(CH_2)_2NH(CH_2)_3NH(CH_2)_2NH_2$) as an alternative to TRIEN. The copper(II) complex with 2,3,2-tetramine is more stable than the equivalent TRIEN complex – cf. $\log \beta = 23.9$ and $\log \beta\, 20.2$, respectively. In addition the former polyamine is easier to purify and is free of toxic isomers. In a comparison of 2,3,2-tetraamine, TRIEN and PEN, 2,3,2-tetraamine proved to be the most effective agent followed by TRIEN and PEN at inducing cupriuresis in the rat.

In NIDM cyclic polyamines may find clinical use as complexing agents for $^{99m}$Tc which when chelated by cyclam is cleared through the liver and kidneys of the unanaesthetized mouse[290]. Jackels et al.[291] have presented data which indicates that manganese(II)-cyclam is more effective than gadolinium(III)-DTPA as a contrast agent.

### 3.5.2 Phosphinepolyacetic Acids

Phosphinotriacetic acid ($P(CH_2COOH)_3$) and ethylenediphosphinetetraacetic acid (EDTPA) are structural analogues of NTA and EDTA[292, 293]. The presence of phosphorus confers considerable softness upon the molecules. As these chelating agents possess both hard and soft donors they could be of use in chelation therapy for toxic metals and NIDM. The chelates of EDPTA have been studied by $^1$H-NMR, $^{31}$P-NMR and UV-visible spectra for a variety of cations. The chelates of Ca(II), Mn(II), Zn(II), La(III) and Pb(II) are of rather limited stability with bonding only through the carboxyl groups. With the exception of Pb(II) and Cd(II), there was no coordination of phosphorus. A rapid intermolecular exchange characterized these complexes. The species present were ML, MHL and $MH_2L$. In contrast the chelates of Fe(II), Co(II), Ni(II), Pd(II), Pt(II) and Hg(II) were stable $ML_2$ complexes which were kinetically inert to intermolecular exchange. The spectra and magnetic measurements indicate bonding of two ligands through four phosphorus atoms. The carboxyl groups are not coordinated and are protonated independently, in agreement with the stability of the complexes in acidic media.

## 3.6 Soft Base Chelating Agents

The soft base chelating agents used in the clinical management of heavy metal poisoning are primarily the 2,3-dimercaptans, PEN, AcPEN, dithiocarbamic acids and thiosemicarbazones. Also included in this category should be placed dicarbothioic acid and xanthates. The acid sensitivity of the latter makes them unsuitable chelating moieties for many applications. The chelating properties of pyridine-2,6-dicarbothioic acid, a fermentation product, have already been observed[294]. No reports have appeared of the use of this chemical in medical sciences.

### 3.6.1 Thiosemicarbazones

The aromatic and heteroaromatic thiosemicarbazones (**39**) are powerful chelating agents which have been rather neglected in the mainstream of chelation chemistry. Perrin and Stünzi[43] have reviewed their role as antiviral agents and noted their use in 1950 to reduce the severity of vaccinia infections of chick embryos and mice. The thiosemicarbazones have been used prophylactically to prevent outbreaks of smallpox in persons who had been in contact with the disease[295]. Pfau[296] has reviewed the pharmaceutical applications of thiosemicarbazones.

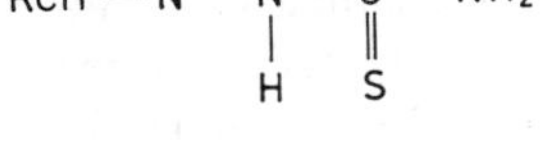

**39**

The chelating properties of the thiosemicarbazones[297)] are regarded as their main chemotherapeutic principle. In the case of the acetylpyridine thiosemicarbazones, only the 2-substituted derivative shows any appreciable antiviral activity. The absence of any significant antiviral activity in the 3- and 5-substituted derivatives supports the hypothesis that a metal chelate is formed between the ring nitrogen and the thiosemicarbazone moiety[298)]. Investigations of the mode of the action of 1-methylisatin β-thiosemicarbazone, the prophylactic used successfully in the treatment of people who had been exposed to smallpox-infected patients, indicate that its copper(II) chelate binds to nucleic acids in vitro under conditions in which there is no significant binding of aquated copper(II) ions[299)]. It is possible that the copper chelate is bound to an mRNA which is synthesized late in the infective cycle. It is also known that the thiosemicarbazone acts only in the final process of infection and is active even after the synthesis of viral DNA has ceased. As the smallpox virus contains appreciable amounts of copper(II), it is possible that its chelation is part of the mode of action.

The other thiosemicarbazones are less well studied and as yet the link between antiviral action and chelation is not fully established. It has been proposed that the chelation of iron(II), a cofactor of ribonucleoside diphosphate reductase, could be the principal mode of action of the thiosemicarbazones[300)]. However, other mechanisms are possible. Investigations of the ESR spectra of copper(II) complexes of thiosemicarbazones has been used to follow the intracellular reactions of the complexes – see Antholine et al.[301)] for a review. In Ehrlich cells the chelate becomes localized in the cell membrane[302)]. This spectroscopic technique could also be used to monitor the antimalarial action of 2-acetylpyridine thiosemicarbazones[303)].

In a recent search for new antitumour chelates of thiosemicarbazones, Mohan et al.[304)] have examined the 4-(m-aminophenyl)-2-formylpyridine thiosemicarbazone chelates of cobalt(II), nickel(II), copper(II), zinc(II) and platinum(II) for antitumour properties but found antitumour activity absent.

The chelating properties of the thiosemicarbazones have not gone unnoticed in nuclear medicine. Arano et al.[305)] have proposed coupling p-carboxyethylphenylglyoxal-di(N-methylthiosemicarbazone) through the p-carboxy moiety to proteins and radiolabelling the immobilized chelating agent with $^{99m}Tc$.

### *3.6.2 2,3-Dimercaptans*

It has already been indicated that BAL has a long history in the treatment of heavy metal poisoning. BAL is an oil-soluble and foul-smelling drug which is unpleasant to take and which is not without serious side-effects. Investigations into other 2,3-dimercaptans have been extensive. Initially the syntheses of polymethyoxydithiols, $HO \cdot CH_2[CH \cdot OH]_n \cdot CH(SH)CH_2SH$, was pursued for alternatives to BAL. In the series $n = 0, 1, 2$ or 3, anti-arsenical activity reached zero when $n = 3$[306)]. For several years now 2,3-dimercapto-propanesulphonic acid (DMPS) and 2,3-dimercaptosuccinic acid (DMSA) have been the alternatives to BAL. In contrast to BAL, both of these chelating agents are less toxic, much more soluble in water, and hence have limited solubility in lipids, and are effective when taken orally. The use of both DMSA and DMPS in combatting heavy metal poisoning has been reviewed by Aposhian in 1983[307)]. Many laboratory studies, reviewed by Aaseth[308)] and Aposhian[307)], have demonstrated the efficacy of DMSA and DMPS in mobilizing inorganic mercury, cadmium, arsenic, copper, lead, gold and antimony.

As yet, neither DMSA nor DMPS have become approved medicines outside the USSR and Eastern Europe. There have, however, been some limited investigations of both chelating agents in mobilizing lead in man. Chisholm and Thomas[309)] have recently administered DMPS to children suffering from lead poisoning. Fractionated doses of DMPS, 200–400 mg per $m^{-2}$ $d^{-1}$, administered over 5 days increased urinary excretion of lead, zinc and copper without any significant changes in hepatic, renal or haematological function. If DMPS lacks the side effects of PEN, it could prove useful in the treatment of chronic lead poisoning in children. Graziano et al.[310)] have observed a similar mobilization of lead after administration of DMSA to eighteen men with elevated blood lead. The efficacy of this therapy can be judged by an increase in red blood cell δ-aminolevulinic acid dehydratase activity and a fall in δ-aminolevulinic acid and coproporphyrin.

As well as the above 2,3-dimercaptoalkanes a few other vicinal dithiols have also been sythesized. The interaction of methylmercury(II) with chelating agents such as BAL involves linear two-coordinate complexes and in only a few cases has chelation involved a bidentate ligand. In a search for evidence of thiol-containing bidentate chelation, Alcock et al.[311)] synthesized three sterically constrained dithiols – cyclohexane-1,2-dithiol, toluene-3,4-dithiol and bicyclo[2.2.1]heptane-2,3-dithiol – and demonstrated by spectroscopy and crystallography their formation of chelates with methylmercury(II) of the form **(40)**.

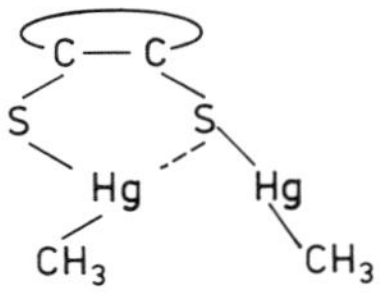


**40**

N-(2,3-Dimercaptopropyl)phthalamidic acid (**41,** DMPA) has been shown to form relatively stable complexes with cadmium, zinc and mercury[312)]. DMPA has also been shown to enhance faecal and urinary excretion of mercury in mice and arsenic in mice and rabbits. For the decorporation of arsenic, taken in as arsine, the administration of 3-(tolylthio)propane-1,2-dithiol **(42)** has been proposed in the USSR[313)].

The preparation of new chelating agents containing vicinal thiols, such as the EDTA-bis-(0-2,3-dimercaptopropyl)ester **(43)**, could lead to new metal-binding ligands which

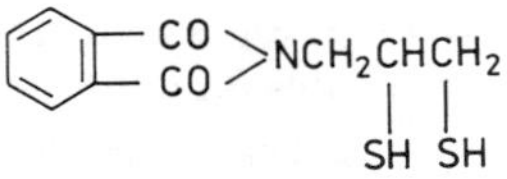


**41**

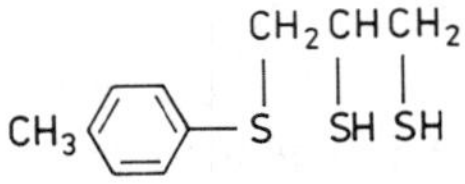


**42**

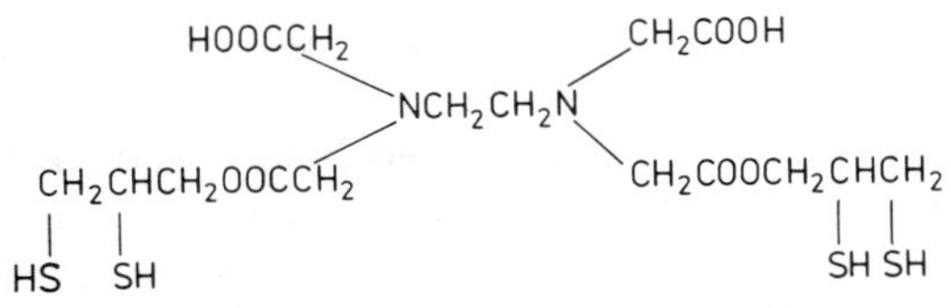


**43**

might have improved chemotherapeutic properties[314]. The preparation of proteins or other macromolecules bearing pendent vicinal thiols could lead to materials with catalytic properties.

In 1974 $^{99m}$Tc-DMSA was proposed as a substitute for organomercurial renal imaging agents[315]. In a subsequent study, an evaluation in mice of $^{99m}$Tc complexed by a series of dimercaptodicarboxylic acids showed the following order of renal clearance: DMSA > 2,4-dimercaptoadipic acid > 2,5-dimercaptoglutaric acid[316]. Investigations of the clearance of $^{99m}$Tc complexed by DMSA emphasize the importance of using $^{99m}$Tc in established oxidation states: $^{99m}$Tc(V) complexed by DMSA has a high uptake into bone whereas a commercially available preparation, marketed as a kidney imaging agent, had a much lower uptake into bone[317]. An alternative renal scanning agent, $^{99m}$Tc-(2,3-dimercaptopropanoic acid), has been recently proposed by Tanaka et al.[318].

## *3.7 D-Penicillamine*

D-Penicillamine, first introduced in 1956 as a therapeutic agent to counter Wilson's disease, has also been used to treat lead poisoning[13]. More recently it has been extensively examined as an agent for the control of rheumatoid disease[13, 319] and as a decuprinating agent in primary biliary cirrhosis[320].

## *3.8 Dithiocarbamic Acid*

Diethyldithiocarbamic acid (DDTC) has been used for some years to combat nickel poisoning arising from intake of nickel carbonyl. The clinical experiences with DDTC have been reviewed by Sunderman[321]. In addition to chelation of nickel(II), DDTC has been examined for ability to mobilize other soft acids. Unfortunately DDTC brings about an unwanted mobilization of cadmium to testes, lung, brain, heart and muscles – see Shinobu et al.[322] for a brief resume of these studies. As this unwanted redistribution can be attributed to the marked lipophilicity of the complex, attention has been directed towards the more water-soluble dithiocarbamates[322]. The compounds synthesized and examined by Jones and his colleagues[322–324] include N-ethyl-N-hydroxyethyldithiocarbamate, N-methyl-N-hydroxyethyldithiocarbamate, N-methyl-D-glucamine dithiocarbamate and the heterocyclic 4-hydroxypiperidine-N-dithiocarboxylate and 4-carboxamidopiperidine-N-dithiocarboxylate – in which the nitrogen atom is contained in a heterocycle. These studies have demonstrated that polar dithiocarbamates do not bring about a redistribution of cadmium to those organs which receive increased cadmiun after therapy with DDTC. In cadmium-loaded mice substantial reductions in both the kidney (71%) and liver (40%) levels were achieved after treatment with N-methyl-D-glucamine dithiocarbamate[322]. The efficacy of these dithiocarbamates in countering the uptake of other soft acids remains undemonstrated. Equally these dithiocarbamates might also serve as alternatives to DDTC in other areas.

Experiments with animals demonstrate that DDTC could be used in conjunction with cisplatin chemotherapy[325, 326]. For instance, pretreatment with DDTC protects against the nephrotoxicity of cisplatin therapy: Wysor et al.[327] have demonstrated the trypanocidal potential of cisplatin and tetraethylthiuram disulphide, the disulphide of DDTC which is reduced to DDTC in the body. Tetraethylthiuram disulphide is marketed as Antabuse, a self-administered drug used by alcoholics to give them an incentive to avoid alcohol. Recent investigations of the metabolism of Antabuse indicate that bis-(diethyl-dithiocarbamate) copper(II) could be formed in plasma[328]. In vitro studies with Antabuse added to plasma showed that $Cu(DDTC)_2$ was slowly formed and that added EDTA competed with DDTC for cupric ions. The low concentrations of $Cu(DDTC)_2$ in patients on Antabuse could be due to the relocation of the complex in fat tissue. The biochemical consequences of deposition of a redox cation in fat tissues remains to be ascertained.

The application of DDTC in nuclear medicine has been examined[329, 330]. DDTC radiolabelled with $^{99m}Tc$ is a non-polar complex with a slow hepatic clearance. Crystallographic investigations of the DDTC chelates of technetium(III) and technetium(V), synthesized by reduction of pertechnetate, show a distorted pentagonal-bipyramidal structure carrying a carbon monoxide ligand for the former produced by reduction by formamidine sulphinic acid, whereas for the latter the chelate is a distorted square-pyramidal structure bearing a nitrido ligand. The carbon monoxide ligand arises from formamidine sulphinic acid and the nitrido ligand is generated by the reductive cleavage of hydrazine. Both of these chelates could have biodistributions quite different from technetium chelates formed in the presence of stannous chloride.

As an alternative to N-isopropyl-p-[$^{123}I$]-iodoamphetamine, an agent used to study regional brain metabolism and perfusion, de Bruin et al.[331] have proposed $^{201}Tl$-DDTC as a replacement. A comparison of radioamphetamine and $^{201}Tl$-DDTC showed that both radiopharmaceuticals are equally suitable for blood flow studies of the brian.

## *3.9 Alkanepolythioethercarboxylic Acids*

A series of thioethercarboxylic acids (**44**, n = 1, 2, 3; **45**) have been synthesized by Jones and his colleagues[332, 333]. These chelating agents have been shown to have a pronounced protective action against methylmercury in mice. These chemicals still await screening in nuclear medicine.

$(HOOCCH_2S)_2CHCH_2CH(SCH_2COOH)_2$

**44**

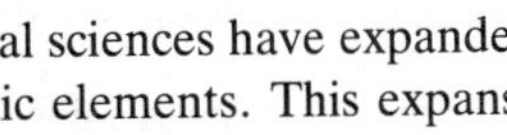

**45**

# 4 Conclusions

The applications of chelating agents in the medical sciences have expanded considerably from their early use in the decorporation of toxic elements. This expansion in use has occurred with improvements in knowledge of inorganic biochemistry.

Two areas for yet further expansion are apparent. Firstly, the continued use of new chelating agents to carry radionuclides, such as $^{99m}Tc$, and paramagnetic nuclei, such as gadolinium(III) and manganese(II), to selected targets in the body. Secondly, the development of new chelating agents to carry metal ions to selected sequences of nucleic acids.

An increased awareness of the organic chemistry of chelating of agents remains a priority in for the continued expansion inorganic biochemistry.

# 5 Abbreviations

| | |
|---|---|
| NMR | nuclear magnetic resonance |
| NIDM | non invasive diagnostic medicine |
| EDTA | ethylenediaminetetraacetic acid |
| DFOA | deferrioxamine |
| cisplatin | cisdiamminodichloroplatinium(II) |
| DNA | deoxyribonucleic acid |
| RNA | ribonucleic acid |
| NTA | nitrilotriacetic acid |
| HEDTA | N-hydroxyethylethylenediaminetetraacetic acid |
| HPDTA | 2-hydroxypropane-1,3-diaminetetraacetic acid |
| PDTA | propane-1,2-diaminetetraacetic acid |
| DCTA | 1,2-diaminocyclohexanetetraacetic acid |
| TAAHA | tris-(2-aminoethyl)aminehexaacetic acid |
| CPDTA | cyclopentane-1,2-diaminotetraacetic acid |
| DTPA | diethylenetriaminepentaacetic acid |
| BADE | 2,2′-bis[di(carboxymethyl)-amino]diethylether |
| BADS | 2,2′-bis[di(carboxymethyl)-amino]diethylsulphide |
| BAETA | 1,2-bis[2-di(carboxymethyl)amino ethylthio]ethane tetraacetic acid |
| TTHA | triethylenetetraaminehexaacetic acid |
| BAL | 2,3-dimercaptopropanol |
| PEN | D-penicillamine |
| AcPEN | N-acetylpenicillamine |
| TRIEN | triethylenetetraamine |
| DMPS | 2,3-dimercaptopropanesulphonic acid |
| DMSA | 2,3-dimercaptosuccinic acid |
| DDTA | N,N-diethyldithiocarbamic acid |
| EHPG | ethylene bis(0-hydroxphenyl)glycine |
| pAPhEDTA | 1-(p-aminophenyl)-ethylenediaminetetraacetic acid |
| 5FBAPTA and Fquene | see basic structures **14** and **15** |
| PE-EDTA | phosphatidylethanolamido-EDTA |
| Puchel-$Cys_2$ | structure **18** |
| EDTA-$Cys_2$ | structure **19** |
| ICRF | Imperial Cancer Research Fund |
| bimolane | see basic structure **22** |
| EDTAHA | ethylenediaminetetraacetohydroxamic acid |
| EDTMP | ethylenediaminetetramethane phosphonic acid |
| DTPMP | diethylenetriaminepentamethylene phosphonic acid |
| MBP, HMBP, HEBP, $Cl_2MBP$, CBP, DCMP, AHPrBP, AHexBP, NMBP, DMBP | see Table 5 |
| ATP | adenosinetriphosphate |
| MCAs | macromolecular chelating agents |
| HSA | human serum albumin |
| BLEDTA | bleomycin bearing pAPhEDTA |
| GI | gastrointestinal |
| CAM | catecholamide |
| CAMC | catecholamide-4-carboxy |
| CAMS | catecholamide-5-sulpho |
| LICAM, LICAMC, LICAMS | linear aminoalkane derivatives such as **30** |
| CYCAM | cyclic catecholamides such as **31** |
| MECAM and TRIMCAM | variants of structure **32** |
| 3,4-DiP-LICAMS | N,N″-bis(isopropyl)-N,N′,N″-tris(5-sulpho-2, 3-dihydroxybenzoyl)-1,5,10-triazadecane] |
| 3,4-DiP-LICAM | N,N″-bis(isopropyl)-N,N′,N″-tris(2,3-dihydroxybenzoyl)-1,5,10-triazadecane |
| TiP-MECAMS | 1,3,5-N,N′,N″-tris(isopropyl)-N,N′,N″-tris(5-sulpho-2,3-dihydroxybenzoyl)triaminomethylbenzene |
| 3,3-LICAMC | N,N′,N″-tris(2,3-dihydroxy-4-carboxybenzoyl)-1,5,10-triazadecane |
| ALA | δ-aminolevulinate synthase |
| AIP | acute intermittent prophyria |
| kryptofix 222 | 4,7,13,16,21,24[hexaoxa]1, 10-diazabicyclo[8,8,8]hexacosane |
| 2,3,2-tetra-amine | 3,7-diazanonane-1,9-diamine |
| cyclam | 1,4,8,11-tetraazacyclotetradecane |
| DMPA | N-(2,3-dimercaptopropyl)-phthalamidic acid |

# 6 References


1. Wood, J. M.: Naturwissenschaften *62,* 357 (1975)
2. Vallee, B. L., Wacker, W. E. C.: in: The Proteins (ed. Neurath, H.), Vol. V, p. 1, New York, Academic Press 1970
3. Blundell, T. L., Jenkins, J. A.: Chem. Soc. Rev. *6,* 139 (1977)
4. Chenoweth, M. B.: Pharmacol. Rev. *8,* 57 (1966)
5. Seven, M. J., Johnson, L. A. (eds.): Metal-Binding in Medicine, Philadelphia, J. B. Lippincott Co. 1960
6. Frederick, J. F. (ed.): Ann. N. Y. Acad. Sci. *88,* 281 (1960)
7. Johnson, L. A., Seven, M. J. (eds.): Fed. Proc. *20,* Suppl. 10, 1 (1961)
8. Pullman, T. N., Lavender, A. R., Forland, M.: Annu. Rev. Med. *14,* 175 (1963)
9. Catsch, A.: Radioactive Metal Mobilization in Medicine, Springfield, Ill., Charles C. Thomas 1964
10. Chenoweth, M. B.: Clin. Pharmacol. Therap. *3,* 365 (1968)
11. Catsch, A., Harmuth-Hoene, A. E., Mellor, D. P.: International Encyclopedia of Pharmacology and therapeutic applications of agents: The Chelation of Metal Heavy Metals (ed. Sartorelli, A. C.), Oxford, Pergamon Press 1979
12. Jones, M. M.:Therapeutic chelating agents, in: Metal Ions in Biological Systems *16,* p. 1 (ed. Sigel, H.), New York, Marcel Dekker, Inc. 1980
13. May, P. M., Bulman, R. A.: Prog. Med. Chem. *20,* 225 (1983)
14. Craven, P. C., Morrelli, H. F.: West. J. Med. *122,* 277 (1975)
15. Soffler, A.: J. Am. Med. Assoc. *233,* 1206 (1975)
16. Bennett, D. D.: Sci. News *127,* 138 (1985)
17. Peters, H. A.: Curr. Ther. Neurol. Dis. 317 (1985–1986)
18. Nieboer, E., Richardson, D. H. S.: Environ. Pollut. (Ser. B) *1,* 3 (1980)
19. Arena, J. M.: Poisoning, p. 369, Springfield, Illinois, Charles C. Thomas 1970
20. Bergsma, D., Cerami, A., Peterson, C. M., Graziano, J. H. (eds.): Iron Metabolism and Thalassemia, New York, Alan R. Liss, Inc. 1976
21. Anon. Lancet *i,* 479 (1978)
22. Cikrt, M., Poňka, P., Nečas, E., Neuwrt, J.: Br. J. Haematol. *45,* 275 (1980)
23. Baker, E., Vitolo, M. L., Webb, J.: Biochem. Pharmacol. *34,* 3011 (1985)
24. Young, S. P., Baker, E., Huehns, E. R.: Br. J. Haematol. *41,* 357 (1979)
25. Lau, E. H., Cerny, E. A., Wright, B. J., Rahman, Y. E.: J. Lab. Clin. Med. *101,* 806 (1983)
26. Pitt, C. G., Gupta, G., Estes, W. E., Rosenkrantz, H., Metterville, J. J., Crumbliss, A. L., Palmer, R. A., Nordquest, K. W., Sprinkle Hardy, K. A., Whitcomb, D. R., Byers, B. R., Arceneaux, J. E. L., Gaines, C. G., Sciortino, C. V.: J. Pharmacol. Exp. Ther. *208,* 12 (1979)
27. Nagara, H., Yajima, K., Suzuki, K.: J. Neuropathol. Exp. Neurol. *40,* 428 (1981)
28. Vaughan, A. T. M., Keeling, A., Yankuba, S. C. S.: Int. J. Appl. Radiat. Isot. *36,* 803 (1985)
29. Neirinckx, R. D.: Chem. Br. 355 (1986)
30. Jones, A. G., Davison, A.: J. Nucl. Med. *23,* 1041 (1982)
31. Pinkerton, T. C., Desilets, C. P., Hoch, D. J., Mikelsons, M. V., Wilson, G. M.: J. Chem. Educ. *62,* 965 (1985)
32. Ketring, A. R., Deutsch, E., Libson, K., Vanderheyden, J.-L., Sodd, V. J., Nishiyama, H., Lukes, S.: J. Nucl. Med. *24,* p. 9 abstr. (1983)
33. Vanderheyden, J.-L., Deutsch, E., Libson, K., Ketring, A. R.: ibid. *24,* p. 9 abstr. (1983)
34. Subramanyam, V., Linder, K. E., Sammartano, M. C., Pendleton, D. B., Delano, M., Liteplo, M., Camin, L. L.: ibid. *24,* p. 9 (abstr.) (1983)
35. Thakur, M. L.: Int. J. Appl. Radiat. Isot. *34,* 617 (1983)
36. Volkert, W. A., Mckenzie, E. H., Hoffman, T. J., Troutner, D. E., Holmes, R. A.: Int. J. Nucl. Med. *11,* 243 (1984)
37. Andrew, E. R.: Acc. Chem. Res. *16,* 116 (1983)
38. Morris, P. G.: Nuclear Magnetic Resonance in Medicine and Biology, Oxford, Clarendon Press 1986
39. Sorenson, J. R. J.: Prog. Med. Chem. *15,* 210 (1978)
40. Sorenson, J. R. J.: Chem. Br. 1110 (1984)

41. DeAlvare, L. R., Goda, K., Kimura, T.: Biochem. Biophys. Res. Commun. *69,* 687 (1976)
42. Weser, U., Riphter, C., Wendel, A., Younef, M.: Bioinorg. Chem. *8,* 201 (1978)
43. Perrin, D. D., Stünzi, H.: Pharm. Ther. *12,* 255 (1981)
44. Hutchinson, D. W.: Antiviral Res. *5,* 193 (1985)
45. Hutchinson, D. W., Naylor, M., Cullis, P. M.: ibid. *5,* 67 (1985)
46. Djordjevic, C., Wampler, G. L.: J. Inorg. Biochem. *25,* 51 (1985)
47. Albert, A.: Selective Toxicity, London, Chapman and Hall 1973[5]
48. Youatt, J.: Aust. J. Biol. Sci. *35,* 565 (1982)
49. Gershon, H., Shanks, L.: Can. J. Microbiol. *27,* 612 (1981)
50. Pijper, P. J.: Synthesis and Antimycoplasmal Activity of Compounds Structurally Related to 2,2′-Bipyridyl. Ph.D. Thesis, Free University Amsterdam 1980
51. Smit, H.: Studies on the Mode of Action of the Copper(I) Complex of 2,9-Dimethyl-1, 10-Phenanthroline on Paracoccus Denitrificans and Mycloplasma Gallisepticum. Ph.D. Thesis, Free University Amsterdam 1982
52. Smit, H., Van Der Goot, H., Nauta, W. T., Pijper, P. P., Balt, S., De Bolstor, M. W. G., Stouthamer, A. H., Verheul, H., Vis, R. D.: Antimicrob. Agents Chemother. *18,* 249 (1982)
53. Smit, H., Van Der Goot, H., Nauta, W. T., Timmerman, H., De Bolster, M. W. G., Stouthamer, A. H., Vis, R. D.: ibid. *21,* 881 (1982)
54. Gaisser, H.-D., v. d. Gout, H., Timmerman, H.: Eur. J. Med. Chem.-Chim. Ther. *20,* 513 (1985)
55. Anon. Lancet *i,* 72 (1971)
56. Damico, R.: US Pat. 358399; Chem. Abstr. *75,* 76623 (1971)
57. Proctor and Gamble Co.: Fr. Demande 2178819; Chem. Abstr. *80,* 96163 (1974)
58. Dobek, A. S., Klayman, D. L., Scovill, J. P., Dickson, E. T. Jr.: Chemotherapy *32,* 25 (1986)
59. Rogers, H. J., Synge, C., Woods, V. E.: Antimicrob. Agents. Chemother. *18,* 63 (1980)
60. Rogers, H. J., Woods, V. E., Synge, C.: J. Gen. Microbiol. *128,* 2389 (1982)
61. Plaha, D. S., Rogers, H. J., Williams, G. W.: J. Antibiot. *37,* 588 (1984)
62. Nabiln, J., El-Wassini, M. T., Karnel, M. M., Metrii, J.: Egypt. J. Bilh. *2,* 177 (1975)
63. Fritsch, G., Freumer, J., Spiro, D. T., Jung, A.: Exp. Parasitol. *60,* 171 (1985)
64. Rosenberg, B.: Clinical Aspects of Platinum Anticancer Drugs, in: Metal Ions in Biological Systems, Vol. 11, p. 1 (ed. Sigel, H.), New York, Marcel Dekker Inc. 1980
65. Sadler, P. J.: Chem. Br. 182 (1982)
66. Cleare, M. J.: in: Structure-Activity Relationships of Anti-tumour Agents (eds. Connors, T. A., Pinedo, H. M., van de Poll, K. W.), p. 59, The Hague 1983
67. Roberts, J., Thompson, A. J.: Prog. Nucleic Acid Res. Mol. Biol. *22,* 71 (1979)
68. Tonew, M., Tonew, E., Schröer, H.-P., Heyn, B.: Zbl. Bkt. Hyg., I. Abt. Orig. A *249,* 296 (1981)
69. Tonew, M., Tonew, E., Schröer, H.-P., Heyn, B.: ibid. A *249,* 421 (1981)
70. Halliwell, B., Gutteridge, J. M. C.: Biochem. J. *219,* 1 (1984)
71. Halliwell, B., Gutteridge, J. M. C.: Phil. Trans. R. Soc. London B. *311,* 659 (1985)
72. Anon. Lancet *i,* 143 (1985)
73. Butler, J., Halliwell, B.: Arch. Biochem. Biophys. *218,* 174 (1982)
74. Buettner, G. R., Oberley, L. W., Leuthauser, S. W. H. C.: Photochem. Photobiol. *28,* 693 (1978)
75. Halliwell, B.: FEBS Lett. *96,* 238 (1978)
76. Gutteridge, J. M. C., Richmond, R., Halliwell, B.: Biochem. J. *184,* 469 (1979)
77. Giordano, N., Fioravanti, A., Sancasiani, S., Marcolongo, R., Borghi, C.: Br. Med. J. *289,* 961 (1984)
78. Blake, D. R., Winyard, P., Lunec, J., Williams, A., Good, P. A., Crewes, S. J., Gutteridge, J. M. C., Rowley, D., Halliwell, B., Cornish, A., Hider, R. C.: Quart. J. Med. *56,* 345 (1985)
79. Prouse, S. J.: cited in Ref. 72
80. Anderson, R. L., Campbell, R. L.: CRC Crit. Rev. Toxicol. *15,* 1 (1985)
81. Nixon, G. A., Belcher, E. V., Niewenkins, R. J.: Toxicol. Appl. Pharmacol. *21,* 244 (1972)
82. Sifri, M., Lowry, D. C., Kratzer, F. H., Norris, L. C.: J. Nutr. *108,* 719 (1978)
83. Cantilena, L. R., Klaasen, C. D.: Toxicol. Appl. Pharmacol. *63,* 344 (1982)
84. Profitt, R. T., Williams, L. E., Presant, C. A., Tin, G. W., Uliana, J. A., Gamble, R. C., Balleschuieler, J. D.: J. Nucl. Med. *24,* 45 (1983)

85. Goddard, J. G., Sweeney, G. D.: Biochem. Pharmacol. *32,* 3879 (1983)
86. Keen, C. L., Saltman, P., Hurley, L. S.: Am. J. Clin. Nutr. *33,* 1789 (1980)
87. Keen, C. L., Lönnerdal, B., Sloan, M. V., Hurley, L. S.: J. Nutr. *10,* 897 (1980)
88. Brownie, C. F., Aronson, A. L.: Toxicol. Appl. Pharmacol. *75,* 167 (1984)
89. Rosenblatt, D. E., Aronson, A. L.: Exp. Mol. Pathol. *28,* 202 (1978)
90. Brownie, C. F., Haluska, M., Aronson, A. L.: Toxicologist *1,* 148 (1981)
91. Duffield, J. R., May, P. M., Williams, D. R.: J. Inorg. Biochem. *20,* 199 (1984)
92. Foley, P.: National Inst. of Occupational Safety and Health/Department of Health, Education and Welfare Press Release on Lead and Chelating Agents, March 1976
93. FDA Memorandum on Penicillamine, Department of Health, Education and Welfare, May 28, 1976
94. Langärd, S., Norseth, T.: in: Handbook on the Toxicology of Metals (eds. Friberg, L., Nordberg, G. F., Vouk, V. B.), p. 383, Amsterdam – New York – Oxford, Elsevier/North Holland Biomedical Press 1979
95. Management of Persons Accidentally Contaminated with Radionuclides: Recommendations of the National Council on Radiation Protection and Measurements, NCRP Report No. 65, p. 146, 1980
96. Robinson, B., Heid, K. R., Aldridge, T. L., Glenn, R. D.: Health Phys. *45,* 911 (1983)
97. Planas-Bohne, F., Olinger, H.: ibid. *31,* 165 (1976)
98. Rahman, Y. E., Rosenthal, M. W., Cerny, E. A.: Science *180,* 300 (1973)
99. Stather, J. W., Smith, H., James, A. C., Rodwell, P.: in: Diagnosis and Treatment of Incorporated Radionuclides, p. 387, Vienna, Internat. Atomic Energy Agency 1976
100. Pecoraro, V. L., Harris, W. R., Carrano, C. J., Raymond, K. N.: Biochemistry *20,* 7033 (1981)
101. Martell, A. E.: in: Developments of Iron Chelators for Clinical Use (eds. Martell, A. E., Anderson, W. F., Badman, D. G.), p. 67, New York, Elsevier North Holland 1981
102. Rajan, K. S., Mainer, S., Rajan, N. L., Davis, J. M.: J. Inorg. Biochem. *14,* 339 (1981)
103. Lauffer, R. B., Greif, W. L., Stark, D. D., Vincent, A. C., Saini, S., Wedeen, U. J., Brady, T. J.: J. Comput. Assist. Tomogr. *9,* 431 (1985)
104. Poldoski, J. E., Bydalek, T. J.: J. Inorg. Nucl. Chem. *41,* 205 (1979)
105. Yeh, S. M., Sherman, D. G., Meares, C. F.: Anal. Biochem. *100,* 152 (1979)
106. Altman, J., Shoef, N., Wilcheck, M., Warshawsky, A.: J. Chem. Soc. Perkin Trans. *I,* 365 (1983)
107. Altman, J., Shoef, N., Wilcheck, M., Warshawsky, A.: ibid. *I,* 59 (1984)
108. Rzessotarski, W. J., Paik, C., Eckelman, W. C., Reba, R. C.: J. Labelled Compd. Radiopharm. *XIII,* 172 (1977)
109. Bard, D. R., Knight, C. G., Page Thomas, D. P.: Clin. Exp. Rheumatol. *1,* 113 (1983)
110. Campbell, A. K.: Intracellular Calcium. John Wiley & Sons Limited. Chichester – New York – Brisbane – Toronto – Singapore 1983
111. Tsien, R. Y.: Nature (London) *290,* 527 (1981)
112. Smith, G. A., Hesketh, T. R., Metcalfe, J. C., Feeney, J., Morris, P. G.: Proc. Natl. Acad. Sci. U.S.A. *80,* 7178 (1983)
113. Rogers, J., Hesketh, T. R., Smith, G. A., Metcalfe, J. C.: J. Biol. Chem. *258,* 5994 (1983)
114. Metcalfe, J. C., Hesketh, J. R., Smith, G. A.: Cell Calcium *6,* 183 (1985)
115. Aiken, J. K.: Chem. Ind. 1334 (1956)
116. Bulman, R. A.: Plzen. Lék. Sborn. Suppl. 49, in press
117. Krescarek, G. E., Tucker, K. L.: Biochem. Biophys. Res. Commun. *77,* 581 (1977)
118. Bersworth, F. C.: French Pat. 1,601,743
119. Alner, D. J., Claret, P. A., Osborne, A. G.: Chem. Ind. 1165 (1968)
120. Markeley, J. F.: Int. J. Rad. Biol. *7,* 405 (1963)
121. Bulman, R. A., Griffin, R. J., Russell, A. T.: Health Phys. *37,* 729 (1979)
122. Bulman, R. A., Griffin, R. J.: Naturwissenschaften *67,* 483 (1981)
123. Volf, V., Peters, E.: Health Phys. *46,* 422 (1984)
124. Crawley, F. E. H., Bulman, R. A., Haines, J. W.: NRPB/R&D3, p. 109 HMSO, London 1978
125. Planas Bohne, F., Lehman, M.: Toxicol. Appl. Pharm. *67,* 408 (1983)

126. Bulman, R. A., Nicholson, J. K., Higham, D. P., Sadler, P. J.: J. Am. Chem. Soc. *106,* 1118 (1984)
127. Nicholson, J. K., Sadler, P. J., Cain, K., Holt, D. E., Webb, M., Hawkes, G. E.: Biochem. J. *211,* 251 (1983)
128. Furst, A.: Chemistry of Chelation in Cancer, Springfield, Illinois, C. C. Thomas 1963
129. Creighton, A. M., Jeffrey, W. A., Long, J.: Proc. 6th Int. Symp. Med. Chem. Chem. Brighton UK, Cotswold Press, Oxford 1979
130. Bakowski, M. T.: Cancer Treat. Rep. *3,* 95 (1976)
131. Huang, Z. X., May, P. M., Quinlan, K. M., Williams, D. R., Creighton, A. M.: Agents Actions *12,* 536 (1982)
132. Camerman, N., Hempel, A.: Science *225,* 1165 (1984)
133. Johansson, R. S., Falch, D. K.: Eur. J. Nucl. Med. *3,* 179 (1978)
134. Karesh, S. M., Eckelman, W. C., Reba, R. C.: J. Pharm. Sci. *66,* 225 (1977)
135. Làzickova, A., Karlicek, R., Kvetina, J., Kronrad, L., Kusak, V.: Int. J. Nucl. Med. Biol. *11,* 170 (1984)
136. Wesley, G. E., Higgins, C. B., McNamara, M. T., Engelstad, B. L., Lipton, M. J., Ehman, R. L., Brasch, R. C., Lovin, J. D.: Radiology *153,* 165 (1984)
137. Carr, D. H., Brown, J., Bydder, F. M., Weinmann, H. J., Speck, U., Thomas, D. J., Young, I. R.: Lancet *i,* 484 (1984)
138. Felix, R., Schoener, W., Claussen, C., Fiegler, W., Neindorf, H. P.: Radiology *153,* 84 (1984)
139. Bydder, G. M., Kingsley, D. P. E., Brown, J., Niendorf, H. P., Young, I. R.: J. Comput. Assit. Tomogr. *9,* No. 4, 690 (1985)
140. Bydder, G. M., Brown, J., Niendorf, H. P., Young, I. R.: ibid. *9,* 847 (1985)
141. Revel, D., Brasch, R. C., Paajanen, H., Rosenau, W., Grodd, W., Engelstad, B., Fox, P., Winkelhake, J.: Radiology *158,* 319 (1986)
142. Runge, V. M., Clanton, J. A., Smith, F. W., Hutchinson, M. J., Partain, C. L., James, A. E.: Roentgenology *141,* 943 (1983)
143. Lovin, J. D., Wesbey, G. E., Engelstad, B. L., Sosnovsky, G., Moseley, M., Tuck, D. L., Brasch, R. C.: Mag. Res. Imag. *3,* 73 (1985)
144. Doughtery, G., Pilbrow, J. R.: Int. J. Biochem. *16,* 1179 (1984)
145. Burger, R. M., Peisach, J., Horwitz, S. B.: Life Sci. *28,* 715 (1981)
146. Marshall, L. E., Graham, D. R., Reich, K. A., Sigman, D. S.: Biochemistry *20,* 244 (1981)
147. Cone, R., Hasan, S. K., Lown, J. W., Morgan, A. R.: Can. J. Biochem. *54,* 219 (1976)
148. Hertzberg, R. P., Dervan, P. B.: J. Am. Chem. Soc. *104,* 313 (1982)
149. Schultz, P. G., Taylor, J. S., Dervan, P. B.: ibid. *104,* 6861 (1982)
150. Schultz, P. G., Dervan, P. B.: Proc. Natl. Acad. Sci. U.S.A. *80,* 6834 (1983)
151. Schultz, P. G., Dervan, P. B.: J. Am. Chem. Soc. *105,* 7748 (1983)
152. Taylor, J. S., Schultz, P. G., Dervan, P. B.: Tetrahedron *40,* 457 (1984)
153. Boutorin, A. S., Vlassov, V. V., Kazakov, S. A., Kutiavin, I. V., Podymonogin, M. A.: FEBS Lett. *172,* 1560 (1984)
154. Kean, J. M., White, S. A., Draper, D. E.: Biochemistry *24,* 5062 (1985)
155. Dreyer, G. B., Dervan, P. B.: Proc. Natl. Acad. Sci. U.S.A. *82,* 968 (1985)
156. Youngquist, R. S., Dervan, P. B.: J. Am. Chem. Soc. *107,* 5528 (1985)
157. Anbar, M., Aviad, I., Rein, R., Schorr, S.: Experientia *16,* 443 (1960)
158. Shtacher, G., Anbar, M.: J. Pharm. Exp. Ther. *152,* 157 (1966)
159. Loberg, M. D., Cooper, M., Harvey, E., Callery, P., Faith, W.: J. Nucl. Med. *17,* 633 (1976)
160. Callery, P. S., Faith, W. C., Loberg, M. D., Fields, A. T., Harvey, E. B., Cooper, M. D.: J. Med. Chem. *19,* 963 (1976)
161. Van Wyck, A. J., Fourie, P. J.,Van Zyl, W. H., Lötter, M. G., Minnaar, P. C.: Eur. J. Nucl. Med. *4,* 445 (1979)
162. Choitellis, E., Varvarigou, A.: Int. J. Nucl. Med. Biol. *7,* 1 (1980)
163. Graham, B. W., Desuliis, C. M., Mattson, R. J., Sowell, J. W.: J. Pharm. Sci. *71,* 362 (1982)
164. Schachner, E. R., Gil, C., Atkins, H. C.: J. Nucl. Med. *20,* 682 (1979)
165. Subramanian, G., McAfee, J. G., Rosenstreich, M., Coco, M.: ibid. *16,* 1080 (1975)
166. Szot, Z., Zabłotna, R., Geisler, J., Poczynajlo, A.: Int. J. Radiat. Biol. *34,* 187 (1978)
167. Szot, S.: EULEP Newsletter, No. 30, 46 (1982)
168. Taliaferro, C. H., Martell, A. E.: J. Coord. Chem. *13,* 249 (1984)

169. Russell, R. G. G., Fleisch, H.: The Biochemistry and Physiology of Bone IV, 61 (1976)
170. Fleisch, H.: in: Bone Resorption, Metastasis and Diphosphonates (ed. Garattini, S.), New York, Raven Press 1985
171. Libson, K., Deutsch, E., Barnett, B. L.: J. Am. Chem. Soc. *102,* 2476 (1980)
172. Russell, R. G. G., Smith, R., Preston, C., Walton, R. J., Woods, G. G.: Lancet *i,* 894 (1974)
173. Instituto Gentil, Pisa SpA: Belg BE 891, 891: Chem. Abstr. *97,* 168913b
174. Edmonds-Alt, X., Brelière, J.-C., Roncucci, R.: Biochem. Pharmacol. *34,* 4043 (1985)
175. Jurisson, S. S., Benedict, J. J., Elder, R. C., Whittle, R., Deutsch, E.: Inorg. Chem. *22,* 1332 (1983)
176. Deutsch, E., Barnett, B. L.: in: Inorganic Chemistry in Biology and Medicine (ed. Martell, A. E.), p. 130, Washington, D.C., Am. Chem. Soc. 1980
177. Francis, M. D., Ferguson, D. L., Tofe, A. J., Bevan, J. A., Michaels, S. E.: J. Nucl. Med. *21,* 1185 (1980)
178. Syhre, R., Johannsen, B., Bunze, R., Mocker, D.: in: Technetium and Nuclear Medicine (eds. Deutsch, E., Nicolini, M., Wagner, H. N.), p. 39, Verona, Cortina-International 1983
179. Subramanian, G., McAfee, J. G., Thomas, F. D., Feld, T. A., Zapf-Longo, C., Palladino, E.: Radiobiology *149,* 823 (1983)
180. Wang, T. S. T., Fawwraz, R. A., Johnson, Z. J., Johnson, P. M.: J. Nucl. Med. *21,* 767 (1980)
181. Wilson, G. M., Pinkerton, T. C.: Anal. Chem. *57,* 246 (1985)
182. Cheng, K. T., Shaw, S. M., Pinkerton, T. C., Hoch, D. J., Van Sickle, D.: Int. J. Nucl. Med. Biol. *12,* 197 (1985)
183. Schümichen, C., Koch, K., Kraus, A., Kuhlicke, G., Weiler, K., Wenn, A., Hoffman, G.: J. Nucl. Med. *21,* 1080 (1980)
184. Jovanović, V., Maksin, T., Rastovac, M., Bzenić, J.: Eur. J. Nucl. Med. *8,* 179 (1983)
185. Savelkoul, T. J. F., Oldenburg, S. J., Van Oort, W. J., Duursma, S. A.: Int. J. Appl. Radiat. Isot. *35,* 709 (1984)
186. Potsaid, M. S., Irwin, R. J., Castronovo, F. P., Prout, G. R., Harvey, W. J., Francis, M. D., Tofe, A. J., Zamenkoff, R. G.: J. Nucl. Med. *19,* 98 (1978)
187. Mathieu, L., Chevalier, P., Galy, G., Berger, M.: Int. J. Appl. Radiat. Isot. *30,* 725 (1979)
188. Oster, Z. H., Som, P., Srivastava, S. C., Fairchild, R. G., Meinken, G. E., Tillman, D. Y., Sacker, D. F., Richards, P., Atkins, H. L., Brill, A. B., Knapp, F. F., Jr., Butler, T. A.: Int. J. Nucl. Med. Biol. *12,* 175 (1985)
189. Roth, H. J., Höfkens, J.: Arch. Pharm. *315,* 131 (1982)
190. Aggett, P. J., Delves, H. T., Harries, J. T., Bangham, A. D.: Biochem. Biophys. Res. Commun. *87,* 513 (1979)
191. Dodin, A., Gellin, J., Guillou, M.: in: Current Chemotherapy, Proc. 10th Internat. Congress Chemotherapy (eds. Siegenthaler, W., Lüthy, R.), p. 155, Amer. Soc. Microbiol., Washington, D.C. 1977
192. Zribi, A., Rachid, M. S. B.: Bull. Soc. Pathol. Exot. *66,* 597 (1973)
193. Scheibel, L. W., Adler, A.: Mol. Pharm. *22,* 140 (1982)
194. Scheibel, L. W., Stanton, G. G.: IUPHAR 9th Internat. Congress Pharmacology Proc. *2,* 385 (1984)
195. Scheibel, L. W., Adler, A.: Mol. Pharm. *20,* 218 (1981)
196. Albert, A., Rees, C. W., Tomlinson, A. J. H.: Br. J. Exp. Pathol. *37,* 500 (1956)
197. Albert, A., Rubbe, S. D., Goldacre, R. J., Balfour, B. G.: ibid. *28,* 69 (1947)
198. Goedemans, W. T., Hardman, M. R., Belfer, A. J.: Eur. J. Nucl. Med. *5,* 63 (1980)
199. Yano, Y., Budinger, T. F., Ebbe, S. N., Mathis, C. A., Singh, M., Brennan, K. M., Moyer, B. R.: J. Nucl. Med. *26,* 1429 (1985)
200. McKillop, J. H., Wallwork, J., Reitz, B. A., Billingham, M. E., Miller, R., McDougal, I. R.: Eur. J. Nucl. Med. *7,* 162 (1982)
201. Anghileri, L. J., Firusian, N., Heidbreder, M.: ibid. *2,* 109 (1977)
202. Hodgkin, J. H.: Chem. Ind. 153 (1979)
203. Kaneko, M., Tsuchida, E.: Macrom. Rev. *16,* 397 (1981)
204. Sahni, S. K., Reedijk, J.: Coord. Chem. Rev. *59,* 1 (1984)
205. Aaseth, J., Eldjam, L., Jellum, S.: Biochem. Pharmacol. *22,* 1179 (1973)
206. Ramirez, R. S., Andrade, J. D.: J. Macromol. Sci. Chem. *A10,* 309 (1976)
207. Dawson, M. I., Hobbs, P. D.: J. Polym. Sci. *21,* 381 (1983)

208. Bulman, R. A., Griffin, R. J.: Health Phys. *40,* 228 (1981)
209. Bulman, R. A., Anderson, O., Nielsen, J. B.: Acta Pharmacol. Toxicol. *58,* (Suppl. 11) 1986
210. Hnatowich, D. J., Layne, W. W., Childs, R. L., Lanteigne, D., Davis, M. A., Griffin, T. W., Doherty, P. W.: Science *220,* 613 (1983)
211. Chanachai, W., Shani, J., Wolf, W., Harwig, J. F., Nakamura, R. M.: Int. J. Nucl. Med. Biol. *12,* 289 (1985)
212. Motta-Hennessy, C., Eccles, S. A., Dean, C., Coghlan, G.: Eur. J. Nucl. Med. *11,* 240 (1985)
213. Hnatowich, D. J., Griffin, T. W., Koscinczyk, C., Rusckowski, M., Childs, R. L., Mattis, J. A., Slealy, D., Doherty, P. W.: J. Nucl. Med. *26,* 849 (1985)
214. Sundberg, M. W., Meares, C. F., Goodwin, D. A., Diamanti, C. I.: Nature (London) *250,* 587 (1974)
215. Janok, G. Y. A., Harwig, J. F., Chanachai, W., Wolf, W.: Int. Appl. Radiat. Isot. *34,* 871 (1983)
216. Winston, A., Kirchner, D.: Macromolecules *11,* 597 (1978)
217. Varaprasad, D. V. P. R., Rosthauser, J., Winston, A.: J. Polym. Sci. *22,* 2131 (1984)
218. Winston, A., Varaprasad, D. V. P. R., Metterville, J. J., Rosenkrantz, H.: J. Pharm. Exp. Ther. *232,* 644 (1985)
219. DeRiemer, L. H., Meares, C. F., Goodwin, D. A., Diamanti, C. I.: J. Med. Chem. *22,* 1019 (1979)
220. DeReimer, L. H., Meares, C. F., Goodwin, D. A., Diamanti, C. I.: J. Labelled Compd. Radiopharm. *XVIII,* 1517 (1980)
221. Buchstaum, D. J., Hanna, D. E., Randall, B. C., Buchegger, F., Mach, J.-P.: Int. J. Nucl. Med. Biol. *12,* 79 (1985)
221a. Vaughan, A. T. M., Bradwell, A. P., Dykes, P. W., Anderson, P.: Lancet *i,* 1492 (1986)
222. Brady, T. J., Rosen, B. R., Gold, H. K., Khaur, B. A., Fallong, J. T., Goldman, H. R., Ter Penning, B. J., Yasuda, T., Haber, E.: Work in Progress Annual Meet. of Soc. of Magnetic Resonance in Medicine, San Francisco, August 1983. Abstr. 10
223. Reardan, D. T., Meares, C. F., Goodwin, D. A., McTigue, M., David, G. S., Stone, M. R., Leung, J. P., Bartholomew, R. M., Frincke, J. M.: Nature (London) *316,* 265 (1985)
224. Dubrovina, Z. U., Dol'matov, M. Yu, Malkin, P. M., Sharonov, G. E., Panteleeva, A. P.: Gig. Sanit. (Engl. Transl.) *34,* 286 (1969)
225. Bulman, R. A., Vanderborght, O., Van Puymbroeck, S.: Health Phys. *44,* 428 (1983)
226. Bakir, F., Damlusi, S. F., Amin-Zaki, L., Murtadha, M., Khalidi, A., Al-Rauri, N. Y., Tikriti, S., Dhahir, H. I., Clarkson, T. W., Smith, J. C., Doherty, R. A.: Science *181,* 230 (1973)
227. Durbin, P. W., Jeung, N., Jones, E. S., Weitl, F. L., Raymond, K. N.: Radiat. Res. *99,* 85 (1984)
228. Schloss, J. V., Smith, G., Aulabaugh, A., Cleland, W. W.: Anal. Biochem. *120,* 176 (1982)
229. Barnes, R. M.: Biol. Trace Elem. Res. *6,* 93 (1984)
230. Porath, J., Olin, B.: Biochemistry *22,* 1621 (1983)
231. Moroux, Y., Boschetti, E., Egly, J. M.: Science Tools (LKB Instr. Journ.) *32,* 1 (1985)
232. Hemmilä, I., Soini, E., Lövgren, T.: Fresenius Z. Anal. Chem. *311,* 357 (1982)
233. Eskola, J.-V., Nevalainen, T.-J., Lövgren, T.: Clin. Chem. *29,* 1777 (1983)
234. Bertoft, E., Eskola, J.-V., Näntö, V., Lövgren, T.: FEBS Lett. *173,* 213 (1984)
235. Ius, A., Ferrara, L., Meroni, G., Bacigalupo, M.: Fresenius Z. Anal. Chem. *322,* 509 (1985)
236. Neilands, J. B., Hider, R. C., Chimiak, A., Liu, A., Sugiura, Y., Nomoto, K.: Struct. Bonding *58,* 1 (1984)
237. Raymond, K. N., Ching, T. D. Y., Pecararo, V. L., Carrano, C. J.: in: The Biochemistry and Physiology of Iron (eds. Saltmann, P., Hegenauer, J.), p. 650. New York – Amsterdam – Oxford, Elsevier Biomedical 1981
238. Raymond, K. N., Pecararo, V. L., Weitl, F. L.: in: Development of Iron Chelators for Clinical Use (eds. Anderson, M., Badman, D. G.), p. 165. Amsterdam, Elsevier N. Holland 1981
239. Raymond, K. N., Tufano, T. P.: in: The Biological Chemistry of Iron, (eds. Dunford, H. B., Dolfin, D., Raymond, K. N., Seiker, L.), p. 85, Dordreck, Holland, D. Reidel Publishing Company 1982
240. Venuti, M. C., Rastetter, W. H., Neilands, J. B.: J. Med. Chem. *22,* 123 (1979)
241. Kontoghiorghes, G. J., Evans, R. W.: FEBS Lett. *189,* 141 (1985)

242. Kontoghiorghes, G. J.: Biochem. J. *233,* 299 (1986)
243. Brooks, H. A., Gardner, D., Poyser, J. P.: J. Antibiot. *XXXVII,* 1501 (1984)
244. Pippard, M. J., Callender, S. T., Weatherall, D. J.: Clin. Sci. Mol. Med. *54,* 99 (1978)
245. Young, S. P., Baker, E., Huehns, E. R.: Br. J. Haematol. *41,* 357 (1979)
246. Green, R., Lamon, J., Curran, D.: Lancet *ii,* 327 (1980)
247. Meyer-Brunot, H. G., Kerberle, H.: Am. J. Physiol. *214,* 1193 (1968)
248. Rodgers, S. J., Raymond, K. N.: J. Med. Chem. *26,* 439 (1983)
249. Grady, R. W., Peterson, C. M., Jones, R. L., Graziano, J. H., Bhargara, K. K., Berdoukas, V. A., Kokkini, G., Loukopoulos, D., Cerami, A.: J. Pharmacol. Exp. Ther. *209,* 342 (1979)
250. Yunice, A. A., Perry, E. F., Perry, H. M.: Arch. Environ. Health *16,* 163 (1968)
251. Ackrill, P., Ralston, A. J., Day, J. P., Hodge, K. C.: Lancet *ii,* 692 (1980)
252. Arze, R. S., Parkinson, J. S., Cartlidge, N. E. F., Britton, P., Ward, M. K.: Lancet *ii,* 116 (1981)
253. Brown, D. J., Dawborn, J. K., Ham, K. N., Xipell, J. M.: Lancet *ii,* 343 (1982)
254. Day, J. P., Ackrill, P., Garstang, F. M., Hodge, K. C., O'Hara, M., Benzo, Z., Ramero-Martinez, R. A.: in: Chemical Toxicology and Clinical Chemistry of Metals (eds. Brown, S. S., Savory, J.), p. 353, London, Academic Press 1983
255. Premuzic, E. T., Francis, A. J., Lin, M., Schubert, J.: Arch. Environ. Contam. Toxicol. *14,* 759 (1985)
256. Rastetter, W. H., Erickson, T. J., Venuti, M. C.: J. Org. Chem. *46,* 3579 (1981)
257. Weitl, F. L., Harris, W. R., Raymond, K. N.: J. Med. Chem. *22,* 1281 (1979)
258. Ardeef, A., Sofen, S. R., Bregante, T. L., Raymond, K. N.: J. Am. Chem. Soc. *100,* 5362 (1978)
259. Weitl, F. L., Raymond, K. N., Smith, W. L., Howard, T. R.: ibid. *100,* 1170 (1978)
260. Weitl, F. L., Raymond, K. N., Durbin, P. W.: J. Med. Chem. *24,* 203 (1981)
261. Weitl, F. L., Raymond, K. N.: J. Am. Chem. Soc. *101,* 2728 (1979)
262. Weitl, F. L., Raymond, K. N.: J. Org. Chem. *46,* 5234 (1981)
263. O'Brien, I. G., Cox, G. B., Gibson, F.: Biochim. Biophys. Acta *237,* 537 (1971)
264. Scarrow, R. C., White, D. L., Raymond, K. N.: J. Am. Chem. Soc. *107,* 6540 (1985)
265. Raymond, K. N., Kappel, M. J., Pecararo, V. L., Harris, W. R., Carano, C. J., Weitl, F. L., Durbin, D. W.: in: Actinides in Perspective (ed. Edelstein, N. M.), p. 491, Oxford – New York, Pergamon Press 1982
266. Lloyd, R. D., Bruenger, F. W., Mays, C. W., Atherton, D. R., Jones, C. W., Taylor, G. N., Stevens, W.: Radiat. Res. *99,* 106 (1984)
267. Stradling, G. N., Stather, J. W., Gray, S. A., Moody, J. C., Ellender, M., Hodgson, A.: Human Toxicol. *5,* 77 (1986)
268. Kappel, M. J., Nitsche, H., Raymond, K. N.: Inorg. Chem. *24,* 605 (1985)
269. Kontoghiorghes, G. J.: The Design of Orally Active Iron Chelators for the Treatment of Thalassaemia. Ph.D. Thesis, University of Essex 1982
270. Koizumi, K., Tonami, N., Hisada, K.: Eur. J. Nucl. Med. *7,* 229 (1982)
271. Herscheid, J. D. M., Hoekstra, A., Vos, C. M.: ibid. *9,* 508 (1984)
272. Moerlein, S. M., Welch, M. J., Raymond, K. N., Weitl, F. L.: J. Nucl. Med. *22,* 710 (1981)
273. Moerlein, S. M., Welch, M. J., Raymond, K. N.: ibid. *23,* 501 (1982)
274. Ovchinnikov, Yu. A., Ivanov, V. T., Shkrob, A. M.: Membrane-Active Complexones, Amsterdam – Oxford – New York, Elsevier 1974
275. Lehn, J.-M.: Struct. Bonding *16,* 1 (1973)
276. Tso, W.-W., Fung, W. F.: Inorg. Chim. Acta *55,* 129 (1981)
277. Arktes, B., King, K., Anderson, R., Peterson, W.: Organometallics *2,* 454 (1982)
278. Pannell, K. H., La Neave, C. K., Rico, E., Arkles, B.: Pharmacol. Biochem. Behav. *21,* 77 (1984)
279. Eaborn, C. A.: Organosilicon Compounds, Chpts. 3 and 9, New York, Academic Press 1960
280. Müller, W. M.: in: Radiation Protection, Proc. 5th Congr. International Radiation Protection Soc., Jerusalem, March 1980, p. 1056, Oxford – New York – Toronto – Sydney – Paris – Frankfurt, Pergamon Press 1980
281. Delgado, R., Frausto Da Silva, J. J. R.: Talanta *29,* 815 (1982)
282. Sun, Y., Martell, A. E., Motekaitis, R. J.: Inorg. Chem. *24,* 4343 (1985)

283. Baldwin, J. E., Perlmutter, P.: Bridged, Capped and Fenced Porphyrins, in: Topics Current Chem. *121,* Host Guest Complex Chemistry III (eds. Vögtle, F., Weber, E.), Berlin – Heidelberg – New York – Tokyo, Springer 1984
284. Kappas, A., Drummod, G. S.: Bioessays *3,* 256 (1985)
285. Vaum, R., Heindel, N. D., Burns, H. D., Enrich, J., Foster, N.: Pharmaceut. Sci. *71,* 1223 (1982)
286. Rousseau, J., Ali, H., Lamoureux, G., Lebel, E., Van Lier, J. E.: Int. J. Appl. Radiat. Isot. *36,* 709 (1985)
287. Walshe, J. M.: Lancet *i,* 643 (1982)
288. Borthwick, T. R., Benson, G. D., Schugar, H. J.: J. Lab. Clin. Med. *95,* 575 (1980)
289. Borthwick, T. R., Benson, G. D., Schugar, H. J.: Proc. Soc. Exp. Biol. Med. *162,* 227 (1979)
290. Troutner, D. E., Simon, J., Ketring, A. R., Volkert, W., Holmes, R. A.: J. Nucl. Med. *21,* 433 (1980)
291. Jackels, S. C., Kroos, B. R., Hinson, W. H., Karstaedt, N., Noran, R. R.: Radiology *159,* 525 (1986)
292. Tzschach, A., Friebe, S.: Zeitschrift Chem. *19,* 375 (1979)
293. Podlahová, J., Šilha, J., Podlaha, J.: Coll. Czech. Chem. Commun. *50,* 445 (1983)
294. Budzikiewicz, H., Ockels, W., Roemer, A., Korth, H., Pulverer, G.: in: Symposium Papers, IUPAC Int. Symp. Chem. Nat. Prod. 11th (Part 2), p. 194 (eds. Marekov, N., Ognyanov, I., Orahovats, A.), Izd BAN, Sofia 1978
295. Bauer, D., St. Vincent, C., Kempe, C., Dounie, A.: Lancet *ii,* 494 (1963)
296. Pfau, C. J.: Handb. Exp. Pharmacol. *61,* 147 (1982)
297. Campbell, M. J. M.: Coord. Chem. Rev. *15,* 279 (1975)
298. Oxford, J. S., Perrin, D. D.: J. Gen. Virol. *23,* 59 (1974)
299. Levinson, W., Rhode, W., Mikelens, P., Antony, A., Ramakuskna, T.: Ann. N. Y. Acad. Sci. *284,* 525 (1977)
300. Auld, D. S., Kawaguchi, H., Livingstone, D. M., Vallee, B. L.: Proc. Natl. Acad. Sci. U.S.A. *71,* 2091 (1974)
301. Antholine, W. E., Kalyanaraman, B., Petering, D. H.: Environ. Health Perspect. *64,* 19 (1985)
302. Campbell, M. J. M., Collis, A. J., Grzeskowiak, R.: Bioinorg. Chem. *6,* 305 (1976)
303. Klayman, D. L., Scovill, J. P., Bartosevich, J. F., Bruce, J.: J. Med. Chem. *26,* 35 (1983)
304. Mohan, M., Sharma, P., Jha, N. K.: Inorg. Chim. Acta *107,* 91 (1985)
305. Arano, Y., Yokoyama, A., Magata, Y., Saji, H., Horiuchi, K., Torizuka, K.: Int. J. Nucl. Med. Biol. *12,* 429 (1986)
306. Miles, L. W. C., Owen, L. N.: J. Chem. Soc. 2934 (1950)
307. Aposhian, H. V.: Annu. Rev. Pharmacol. Toxicol. *23,* 193 (1983)
308. Aaseth, J.: Human Toxicol. *2,* 257 (1983)
309. Chisolm, J. J., Jr., Thomas, D. J.: J. Pharmacol. Exp. Ther. *235,* 665 (1985)
310. Graziano, J. H., Sins, E. S., Lojacono, B. S., Silverberg, S. J., Turgcon, L.: Clin. Pharmacol. Ther. *37,* 431 (1985)
311. Alcock, N. W., Lampe, P. A., Moore, P.: J. Chem. Soc. Dalton Trans., 1471 (1980)
312. Stine, E. R., Hsu, C.-A., Hoover, T. D., Aposhian, H. V., Carter, D. E.: Toxicol. Appl. Pharmacol. *75,* 329 (1984)
313. Mizyukova, I. G., Petrunkin, V. E., Lysenko, N. M.: Farmakol. Toksikol. (Moscow) *34,* 70 (1971); Chem. Abstr. *74,* 97223a
314. Bulman, R. A.: Acta Pharamcol. Toxicol. *58,* Suppl. 11, 553 (1986)
315. Lin, T. H., Khentigan, A., Winchell, H. S.: J. Nucl. Med. *15,* 34 (1974)
316. Kubiatowicz, D. D., Bolles, T. F., Nora, J. C., Ithakissios, D. S.: J. Pharm. Sci. *68,* 621 (1979)
317. Yokoyama, A., Hata, N., Horiuchi, K., Masuda, H., Saji, H., Ohta, H., Yamamoto, K., Endo, K., Torizuka, K.: Int. J. Nucl. Med. Biol. *12,* 273 (1985)
318. Tanaka, A., Machida, T., Miki, M., Shimada, T.: Eur. J. Nucl. Med. *8,* 317 (1983)
319. Munthe, E. (ed.): Penicillamine Research in Rheumatoid Disease: Symposium held at Spåtin, Norway: March 7–10, 1976
320. Jain, S., Scheurer, P. J., Samourian, S., McGee, J. O'D., Sherlock, S.: Lancet *i,* 831 (1977)
321. Sunderman, F. W.: Ann. Clin. Lab. Sci. *11,* 1 (1981)
322. Shinobu, L. A., Jones, S. G., Jones, M. M.: Acta Pharmacol. Toxicol. *54,* 189 (1984)

323. Jones, M. M., Burka, L. T., Hunter, M. E., Basinger, M., Campo, G., Weaver, A. D.: J. Inorg. Nucl. Chem. *42,* 775 (1980)
324. Gale, G. R., Atkins, L. M., Walker, E. M., Jr., Smith, A. B., Jones, M. M., Hodge, R. P.: Res. Commun. Chem. Pathol. Pharmacol. *43,* 281 (1984)
325. Borch, R. F.: Proc. Natl. Acad. Sci. U.S.A. *77,* 5441 (1980)
326. Walker, E. M., Jr., Greene, W. B., Gale, G. R., Nicholson, J. H., Vesely, J., Henniger, G. R.: in: Chemical Toxicology and Clinical Chemistry of Metals (eds. Brown, S. S., Savory, J.), p. 395–8, London, Academic Press 1983
327. Wysor, M. S., Zwelling, L. A., Sanders, J. E., Grenan, M. M.: Science *217,* 454 (1982)
328. Johansson, B., Stankiewicz, Z.: Biochem. Pharmacol. *34,* 2989 (1985)
329. Pojer, P. M., Baldas, J.: Int. J. Nucl. Med. Biol. *8,* 112 (1981)
330. Baldas, J., Bonnyman, J., Pojer, P. M., Williams, G. A.: Eur. J. Nucl. Med. *7,* 187 (1982)
331. de Bruine, J. F., van Royen, E. A., Vyth, A., de Jong, J. M. B. V., van der Schoot, J. B.: J. Nucl. Med. *26,* 925 (1985)
332. Jones, M. M., Harbison, R. D.: Res. Commun. Chem. Path. Pharmacol. *7,* 389 (1974)
333. Jones, M. M., Banks, A. J., Brown, C. H.: J. Inorg. Nucl. Chem. *37,* 761 (1975)

## Author Index Volumes 1–67